W9-CSG-368

Star Crossed

UNIVERSITY PRESS OF FLORIDA

Florida A&M University, Tallahassee
Florida Atlantic University, Boca Raton
Florida Gulf Coast University, Ft. Myers
Florida International University, Miami
Florida State University, Tallahassee
New College of Florida, Sarasota
University of Central Florida, Orlando
University of Florida, Gainesville
University of North Florida, Jacksonville
University of South Florida, Tampa
University of West Florida, Pensacola

S T A R

★ ★

CROSSED

The Story of Astronaut Lisa Nowak

KIMBERLY C. MOORE

University Press of Florida

Gainesville · Tallahassee · Tampa · Boca Raton

Pensacola · Orlando · Miami · Jacksonville · Ft. Myers · Sarasota

Copyright 2020 by Kimberly C. Moore
All rights reserved
Published in the United States of America

25 24 23 22 21 20   6 5 4 3 2 1

Library of Congress Control Number: 2019954699
ISBN 978-0-8130-6654-7

The University Press of Florida is the scholarly publishing agency for the State University System of Florida, comprising Florida A&M University, Florida Atlantic University, Florida Gulf Coast University, Florida International University, Florida State University, New College of Florida, University of Central Florida, University of Florida, University of North Florida, University of South Florida, and University of West Florida.

University Press of Florida
2046 NE Waldo Road
Suite 2100
Gainesville, FL 32609
http://upress.ufl.edu

This book is dedicated to the tens of thousands of people who work for the National Aeronautics and Space Administration and have dedicated their lives to the study and exploration of space, and making life better here on Earth.

Contents

Prologue

Star Crossed is a work culled from thousands of hours of direct reporting by me as I covered Lisa Nowak's criminal case in Orange County, Florida, read thousands of pages of NASA documents and reports, sifted through multiple legal filings, and watched dozens of hours of television news documentaries, NASA videos, court videos, and several police videos. Information also comes from several books, including *Choosing the Right Stuff: The Psychological Selection of Astronauts and Cosmonauts,* by Dr. Patricia Santy; *Magnificent Desolation: The Long Journey Home from the Moon* and *No Dream Is Too High,* by Colonel Buzz Aldrin; and *Gabby: A Story of Courage, Love and Resilience,* by Captain Mark Kelly and U.S. Representative Gabrielle Giffords (both retired).

The chapters in which Nowak is questioned by Orlando Police detective Chris Becton come from transcripts of his interrogation, along with interviews of him. In an attempt to make it orderly, the retelling of her questioning is abbreviated and organized by subject matter, rather than a straight, word-for-word, minute-by-minute dialogue of the six hours of her questioning.

In some instances, I've used information provided by Nowak's friends, who asked to remain anonymous.

I reached out to every astronaut from her flight for an interview. Five either declined or did not respond. One has died, although information from a letter he wrote to a judge is used. Through a spokesperson, her family declined to comment. Lisa Nowak; her ex-husband, Richard Nowak; and William Oefelein have never given an interview about the incident.

Star Crossed

1

★   ★

The Drive

Saturday, February 3, 2007

Astronaut Lisa Nowak woke up in her Houston home at the end of a cul-de-sac on Parsley Hawthorne Court that Saturday not intending to launch herself into infamy.

Nowak was alone in the red brick house. Her fourteen-year-old son and five-year-old twin girls were spending the weekend with her husband, who had moved out just before Christmas. Her son's birthday was that coming week.

She looked into the room with her vast collection of stamps for making cards, the stamps with which her girls loved to play. Her crop of four hundred African violet plants and cuttings, many of which she gave away as gifts, were scattered throughout the house, although some were wilting after weeks of neglect. She had propagated them from just eight plants left behind by her friend Dr. Laurel Clark, who had died on board the space shuttle *Columbia*. It was as though, in her mind, the woman's life somehow continued through these plants.

Lining some of her bookshelves were rows of the crime novels she told reporters she liked to read.

She had been thinking about what she needed to do for herself, envisioning it, planning it, for nearly a month—ever since her boyfriend told her he was seeing someone else.

In mid-January, she had logged in to her work computer on the sixth floor of an office building at Johnson Space Center and went to the MapQuest site to get directions from Houston to Orlando.

She had gone to her local Sports Authority and bought a knife and BB pistol, along with ammunition and a can of pepper spray, paying cash for it all.

That week, at her boyfriend's apartment and on his USS *Nimitz* aircraft carrier stationery, she had hastily scribbled his new girlfriend's flight itinerary, her unlisted home phone number that she had found in his phone bill—the number he called repeatedly. Most importantly, she wrote down the hour that this woman, Colleen Shipman, would arrive back in Orlando on a United Airlines jet late Sunday night, February 4.

At home that Saturday morning, Nowak packed her black duffel bag. Like any well-trained National Aeronautics and Space Administration (NASA) employee with an obsessive attention to detail, she had made a list of what she needed: knife, BB pistol, ammunition, hammer, plastic gloves, and a disguise. And cash—a lot of cash to pay for gas and hotels. No need to leave a paper trail of credit card and debit card purchases. She wrote it all down on her astronaut "flight events/history/briefing" stationery.

She got into her husband's blue BMW, with its fading paint, and left her suburban home, the one with the snapdragons planted around the brick mailbox. She headed to her local grocery store, and, at just about 1:00 P.M., she bought a phone card with one hundred minutes, paying $6.50 in cash. She had left her cell phone, with its tracking device, at home.

It was a cool day on the first Saturday in February. She was leaving behind the famed Johnson Space Center, a place where she had spent ten years perfecting maneuvers in space shuttle and space station replicas, and in simulated zero gravity, packed into the bulk of a space suit. She worked with a team of people who, like her, were the most elite professionals in the world.

And like her, they had perfect credentials, perfect track records, and perfect personal lives. At least to someone peering into their tiny capsule, that's how it all seemed.

Her plan was perfect, too. Nothing could go wrong, she thought.

She merged onto I-45 and then onto I-10, to make a turn that would, nearly a thousand miles later, allow the world to watch her tumble from her privileged pedestal.

But for now, her mind was on one thing alone: finding her boyfriend's new love interest so she could, hopefully, get him back.

Nowak pushed the pedal down, knowing she had a fifteen-hour drive ahead of her. Soon the plains of Texas gave way to the marshes of Louisiana as the sun dipped below the horizon behind her. The streetlights flashed one after another overhead through Mississippi, Alabama, and north Florida.

Nowak didn't want to be noticed on this trip, didn't want to stop. She didn't want people to remember seeing her, and she was purposely limiting her contact with anyone. So, rather than utilize rest-stop bathrooms, she would later tell an Orlando Police detective, she pulled to the side of the road, grabbed one of her twin daughters' diapers from a box she had left behind in the car years earlier, slid it underneath her, and relieved herself. In her world, it was normal for astronauts to use diapers—called maximum absorbency garments by NASA—in flight suits and training suits. She put two used diapers in a large trash bag on the floor of the rear seat.

Just after 10:30 that night, she pulled off the interstate at DeFuniak Springs, a speck on the map between Pensacola and Tallahassee. It was near Eglin Air Force Base, where she and her husband had done survival training while in the Navy. She turned into a Days Inn parking lot, grabbed her bags, and checked in under a fake name—Linda Turner—paying a little more than fifty dollars in cash for the stay.

She crawled into the bed in the generic room, number 118, with a polyester, floral spread on the bed and cheap framed prints on the walls.

Her Independence Day flight into space had taken place just seven months before. She lifted off on that blistering afternoon from launchpad 39-B at Kennedy Space Center, heading to where only a few hundred others had been before her. All as hundreds of thousands of people watched from the space center, the beaches, and rivers, and on television.

She had worked all her life to reach that pinnacle. Her parents encouraged her and told her she had to be the best, could never misbehave, and must have perfect grades. They expected perfection from their three daughters, and she worked hard to please them. And she expected perfection from herself.

In high school, she watched as Sally Ride was selected as one of the first American women astronauts. She knew that if she studied her hardest and became the best in everything she tried, including excelling in advanced classes, playing field hockey, and serving on student government, she might make it, too. All it required was perfection on her part.

Her trajectory to astronaut was a textbook example of creating a plan, following it, and achieving it. She graduated as covaledictorian of her 1981 class at C. W. Woodward High School in Rockville, Maryland. That achievement, her stellar grades, and her leadership skills propelled her to an appointment to the U.S. Naval Academy in Annapolis, Maryland. That's where she met her husband, whom she married in 1988. A stint at Johnson Space Center before obtaining her master's degree in 1992 put her on the right track to become the right stuff.

She applied six times and was eventually selected in 1992 as a test pilot in Patuxent River, Maryland, learning to fly more than two dozen high-performing aircraft, even as she became a mother that year.

Finally, in 1996, NASA called her name and gave her a royal-blue astronaut's jumpsuit. She, her husband, and their young son moved from Patuxent River, Maryland, to Houston, Texas. Her husband became a communications specialist at the command center. And she went through grueling training sessions in conditions that would make the average claustrophobic lose their minds, in a bulky, hot flight suit to simulate zero gravity. She trained for different life-and-death scenarios, again and again. She repeated the same maneuvers over and over, day after day, even after her friends perished on board the space shuttle *Columbia*.

Another part of her job was to visit the various space centers around the country, where parts of the shuttle were made, and talk to the workers there. She was away from home quite a bit.

But when she was home, she had the perfect life. Or so it seemed to family, friends, and neighbors. A nearly twenty-year marriage. A beautiful house in a quiet neighborhood. Two sisters and parents who loved her and had encouraged her all her life.

Despite fertility problems, Nowak wanted more children after listening to her son pray for siblings each night. She suffered the devastating heartbreak of two miscarriages before giving birth to twin girls in the fall of 2001.

One morning in February 2003, she watched in horror as the space shuttle *Columbia* disintegrated upon reentering Earth's atmosphere. Her friends, her astronaut classmates, were lost in the early-morning, clear, blue sky. The space shuttle *Columbia* accident brought indescribable grief and pain into her orderly life. This wasn't supposed to happen.

At the time, her husband, Richard, a fellow U.S. Naval Academy graduate, had been deployed by the Navy for America's War on Terror.

And so it fell to her elementary-school-aged son to comfort her. And in turn she comforted her friend's son and helped astronaut Laurel Clark's family deal with the logistics and paperwork that come with the death of a loved one.

Her 2007 journey from Texas to Florida was taking place a few days following the fourth anniversary of that horrible day.

She also watched as the U.S. space program sat at the precipice of dying. It had been up to her and her fellow crewmates to help save the space shuttle program during their July 2006 flight.

Her marriage had been crumbling for years. She and Richard agreed that they would stay married until after her space shuttle flight, but that was delayed because of the accident and safety concerns from 2003 until 2006.

Her star-spangled flight into low Earth orbit was textbook, NASA perfection. Every detail had been planned, followed, and executed flawlessly by Nowak and her crewmates. When she returned from space, she began making plans with the new man in her life.

Neighbors said they had heard the sounds of dishes breaking in November 2006. A month later and just five months after Nowak went into space, Richard moved out.

She waited for years to tell her parents and sisters that her marriage was falling apart. She hadn't said a word to them about her new love.

Now her husband was living elsewhere, and the lawyers were handling things.

The facade of her life was peeling apart, and all this perfection, she learned, has a price.

She had turned to a colleague for support. Billy Oefelein was handsome, smart, and understood the demands of the job. At first they were just friends—both were married to other people. Then, beginning in 2004, during and after a survival training mission in the frigid temperatures of Canada, they began spending more and more time together: riding bikes, training for triathlons, working out in the astronaut gym. And she began thinking about him more and more. His wife filed for divorce in 2005.

When she told her mother that her marriage was over, the older woman stopped talking to her. This wasn't perfection; this wasn't what they demanded and she had always given to them. It would tarnish the family in the eyes of friends and neighbors on their Rockville, Maryland, street, the one lined with trees that turned golden, amber, and ruby in the fall. And it would sully the family's name with St. Elizabeth's Catholic Church. What would everyone think?

She no longer cared what anyone thought. Anyone except Billy. She suspected something was wrong when he hadn't returned her repeated phone calls after he finished his space shuttle flight in December 2006. Then in early January, he told her: there was someone else, and he was in love with her.

She wished him well and asked if they could still train together for an upcoming bike race. Within days of his news, she hopped on a training flight with him. She clung to hope.

At home, she was sick. She couldn't eat—her stomach in knots—she had lost twenty pounds, with her weight dropping to 107 pounds in the span of about six weeks, making the five-foot-four-inch woman look gaunt. She couldn't concentrate and had been stopped by Houston Police three times. She cried a lot and she couldn't sleep, her mind churning over and over the facts of her life and how to possibly fix everything.

At work, they had recently told her she would not be chosen for one of the last space shuttle flights. She would probably never fly in space again after some of her colleagues had described her as "uncooperative," "prickly," and "not a team player."

Her marriage had failed, she had failed in her job by not getting another shuttle flight, and now her relationship with the other man was failing.

That's when she began thinking. And planning.

She hadn't slept much, if at all, that Saturday night, but checked out of the DeFuniak Springs hotel at 10:45 the next morning and began the final leg of her trip—as Bruce Springsteen sang a "broken hero on a last-chance power drive"—to a hotel near the Orlando airport. She would utilize their parking lot, take their free shuttle to the airport, and then lie in wait.

2

★ ★

The Call

BEEP BEEP BEEP BEEP BEEP BEEP

Orlando Police Detective Chris Becton groaned, turned off his beeper, and looked at the clock: 4:40 A.M. His wife, Bridgette, was asleep beside him.

"This better be really good," he grumbled.

The couple had hosted a Super Bowl party the night before and had barely cleaned up before going to bed after midnight.

"Call comm center," flashed on the pager's screen. He picked up the phone and punched in the numbers to his office.

The operator transferred him to Chuck Hengehold, the assistant squad leader.

He had a laugh in his voice. "Are you awake?"

"Whatcha got?"

"You will never believe this," he said, pausing. "We arrested a NASA astronaut for carjacking."

Thoughts ran through his mind: *Am I still asleep? Having one of those weird Freudian dreams? Am I hearing things? Did Hengehold just say he arrested an astronaut?*

"What?"

Hengehold giggled and said, "She is a card-carrying, bona-fide astronaut."

"No way!"

"I'm dead serious."

It seemed that, in his eighteen-year police career, Becton had either been in the middle of unbelievable events or working them. It had been just forty-five days since the drunken son of a United States senator had pushed Becton, forcing him to arrest the young man for assault on a law enforcement officer. There was also the fatal shooting of a deranged man several years earlier—a shooting that earned him the medal of valor and an officer-of-the-year award for saving a colleague's life. And now this.

He got dressed in the dark, cursing under his breath at having to handle something that was potentially enormous. Bridgette rolled over.

"What's going on?"

"An astronaut has been arrested. I don't know how long I'll be, but I'll call when I get a chance."

He knew it would be at least a day before he saw her again. He kissed her and walked through the dark house, heading out the door to his police cruiser.

His drive to work at the airport usually took about twenty minutes, and, despite the clogged Orlando traffic when he was usually heading to the office, he used it as his quiet time. At this hour, it truly was quiet.

"God, give me wisdom to make the decisions I need to make, grant me the courage to apply the wisdom You provide," he prayed that morning as he drove the nearly empty streets.

Then it hit him. This is going to be like O. J. Simpson, with camera crews from around the world and attention like nothing he has ever done before.

"Oh, Lord, please don't let me screw this up."

His cell phone rang as he got close to the office. It was Hengehold again.

"Hey—this gets better," said Hengehold, leaving Becton to wonder momentarily how. "The astronaut and the victim dated the same guy. And he's an astronaut, too."

Better? This couldn't get any worse. Or could it?

"Oh—and there's a knife in the astronaut's bag. And a BB gun. And latex gloves. And garbage bags."

Becton knew, based on everything he had learned so far, that this was going to be the biggest case of his career.

He walked in the door of the precinct, and the office was a beehive. Patrol officers were peppering him with the details of what had gone on when they arrived at the scene, talked with the victim, and arrested the astronaut. Hengehold pulled him aside and gave him a rundown of the night's events.

Becton looked at her NASA ID badge: Lisa Marie Nowak.

His mind flashed back to the Fourth of July, when he took his youngest daughter, ChristianAnn, to Cocoa Beach, to watch her first space shuttle launch up close. Like thousands of others packed onto the sand that day, they cheered when they saw the streak of flame slowly climb the afternoon sky and then, a few seconds later, they heard the rumble of the launch as the sound washed down the beach and rivers. He had read to his daughter the profiles of all the astronauts he found in the local newspaper. He knew Lisa Marie Nowak had been on board.

He walked into his office and sat behind his old, wooden desk that looked like it might have been Joe Friday's during the *Dragnet* era. He reached into the minifridge behind the desk and grabbed an energy drink, popping the top and gulping down the mixture that would keep him awake all morning.

In order to prepare for the next few hours, he would need to know everything possible about Captain Lisa Marie Caputo Nowak. And the best place to find out anything on someone like this was on Google. He turned on his computer, called up the search engine, and typed in her name. NASA had her life story on one webpage.

Lisa had logged 1,500 flight hours in twenty different aircraft. *Wow.* She can fly an F/A-18 fighter jet. *Great. How could he interview a combat-jet pilot?* She has been to electronic warfare school. *Just dandy.* He read on. She can fly the EA-7L, a snub-nosed jet.

Becton had a flashback to the movie *GI Jane* with Demi Moore. He thought about the school she'd been sent to, where she learned escape, evasion, and interrogation techniques. It's training all military combat personnel go through.

How does a detective interview a combat-jet pilot, get her to give a statement, without coercing the statement, and not violate her civil rights? He thinks, *There is always the go-into-the-interview-stern-and-cranky approach.* Right, like the military hasn't taught her how to deal with a mean interrogator. Or he could go into the interview and be submissive. As he sat thinking in his office, Becton decided the best way to approach the interview with this astronaut was to be up-front and honest. No games. *Miranda is first—got to read the Miranda rights to her first.* He was prepared for her to ask for a lawyer. Anyone who has been trained as well as Lisa had been was going to ask for a lawyer.

But first, he needed to talk to the victim and make sure her story checked out.

3

★　★

Childhood

Lisa Marie Caputo was born in May 1963, the first of three children, to Jane and Alfred Caputo. Two younger sisters, Andrea and Marisa, would follow. Three years later, her parents enrolled her in a nursery school near their Rockville, Maryland, home, nestled in a neighborhood of upper-middle-class homes.

Lisa walked up the sidewalk to her school, to a world of crayons, building blocks, and Dick and Jane stories. She had big blue eyes, and her mother kept her brown straight hair shoulder length, with bangs almost down to her eyebrows. She wore jumpers with turtlenecks and tights underneath.

Each day after school, Lisa would tell her mother, in detail, everything that had happened at school. But after one month, Mrs. Caputo, a biologist, was astounded when the school called her with some news.

"We think you should evaluate Lisa for deafness," an administrator told her mother. "She has not spoken a word to, nor interacted with, any of the teachers or toddlers and prefers to sit by herself, off to the side, and watch what is happening."

"That can't be," her mother said. They invited her to come to the school and observe Lisa in the classroom. She sat where Lisa wouldn't see her and saw for herself that Lisa was, essentially, mute.

She came out from her hiding place and talked with her daughter.

"Why won't you speak to your teachers or the other children?"

Her answer was a little dumbfounding.

"You told me never to talk to strangers," said the quiet child.

Her mother assured her that it was OK to talk to these people, that they were her friends and teachers—not strangers. She immediately began interacting with her young classmates.

The next year in preschool, there was another incident. Lisa, four, was stubbornly adamant in refusing to be disciplined for leaving grape stems on the floor of the lunchroom.

"The stems belonged to a classmate," Lisa recalled decades later. "If my teacher had asked me to pick up my classmate's stems and throw them away, I would have done it without question. But I refused to pick them up because the teacher wrongfully said they belonged to me."

Her parents also wondered why the quiet introvert didn't seek out their hugs, cuddles, or kisses. But, they observed, she didn't reject them when given, either.

A family snapshot and a school photograph both show her looking straight into the camera—but unsmiling. Not pouting or frowning. Just not smiling. By middle school, though, she had it figured out. And in high school, her smile was genuine and warm.

In 2007, Jane and Alfred Caputo, along with Lisa's two younger sisters, would tell psychiatrist Dr. George Leventon about these incidents and other behavioral quirks. They described Lisa as being "emotionally neutral, but a very determined and highly focused" little girl. Andrea described Lisa as "the perfect child, prudish, generally solitary, and always involved in schoolwork or projects." And, her family said, she had always had difficulty picking up on social cues.

Lisa Marie Caputo's entrée into adulthood began early on the dazzling spring morning of June 12, 1981. At her home in the Washington, D.C., suburb, she stood in front of the mirror, brushing her long, silky hair. Her unblemished face stared back at her in the mirror.

This was her day—graduation day from Charles W. Woodward High School. And she had earned the top spot of her class: the coveted honor of being valedictorian, which she shared with classmate Margaret Flather.

She, her parents, and two sisters loaded into the car for the half-hour drive into the city, to the giant "people's shrine to culture." Her mother, Jane, a biological specialist, and father, Alfred, a computer consultant, reassured her that her short speech would be fine.

She was a bright quasar among her classmates, orbiting on a higher academic plane than most. She tackled a series of honors and advanced-placement classes during her four years at the school, including calculus in her senior year. And she maintained the perfect grades for which she had become known among her teachers and classmates. Her peers saw her as diligent, a girl who had earned the right to stand before them that day.

They remembered her as a classmate who loved to learn. Not just for school, but for herself. She researched things not assigned in class, scribbling notes in a thick denim-blue organizational binder.

And she was an athlete, moving with grace on the school's track. She did her best not only to win, but to beat her own time.

Her friend Alison Ahmed caught a glimpse of things to come. She told *Newsweek* magazine in 2009 that during field-hockey practice, as the team was running laps, Lisa was always out front—and alone.

"Lisa had to be first," Ahmed said.

She sparkled when she sat on the grass, laughing with friends, her nose crinkling a little, her eyes glittering. These were happy times for an all-American girl, the kind of girl every mother hopes to have—or wants their son to marry. She had a boyfriend when she was in the tenth grade, a relationship she said was innocent. Her family had no idea that she considered the boy to be anything more than a friend. She shared no secrets with her younger sisters.

She was never in trouble. Never had a bad grade. Her dreams of becoming a midshipman in Annapolis were coming true. She had visited the Naval Academy the year before and liked what she saw. She was also honored to be the first person in her family to serve in the military, and that was part of what attracted her to the academy.

Her aspirations to fly for the Navy were closer than a dream. And the astronaut corps . . . she was hoping she could join the ranks of Sally Ride, who was training to go on her first mission that year.

Her classmates liked Lisa. As a matter of fact, no one could imagine anyone disliking her. They thought she was nice, unassuming, and approachable. They listened to Kim Carnes and REO Speedwagon on the radio and watched *The Blues Brothers* and *The Empire Strikes Back* at the movie theater together. She was someone any girl or boy could turn to as a friend at the lunchroom table or share a Slurpee with on the football field behind the school.

Her friend Claudia Kalb wrote in a *Newsweek* magazine article that Lisa Caputo was "hyper-focused, not teenaged silly."

At Kalb's Sweet-16 birthday party, she said Lisa looked "angelic at the breakfast table in her pink robe."

But Kalb said that, despite taking classes together, despite taking piano lessons together, despite celebrating birthdays and posing for snapshots together . . . Lisa never once confided in her.

"Nor did she take a huge amount of joy in the closeness and wonder of teenage friendship," Kalb said.

Woodward High School sat in an upper-middle-class area of Montgomery County, a place where the children of U.S. senators and rich businessmen could attend without the added burden of tuition, but still be assured of receiving the same good education as the boys going to neighboring Georgetown Prep, where future U.S. Supreme Court justice Brett Kavanaugh was attending as a sophomore that year.

Outside their cocoon in Rockville, though, times were turbulent. Just a few months before, someone had tried to assassinate President Reagan. Someone did kill John Lennon the previous December. And fifty-two American hostages had been released in January from Iran, after more than a year in captivity at the American embassy.

On April 12, Lisa had watched the news as a new type of spaceship launched into orbit. It went up like a rocket but looked like an airplane. The space shuttle *Columbia* made the fleet's maiden trip into space, proving that the vehicle could go up, operate in orbit, and land safely.

Lisa Caputo was graduating, and the local civic center wouldn't be good enough for the 313 Wildcats picking up their diplomas that morning. Driving down Canal Road, peering through the overhanging trees bright with Crayola-green leaves, Lisa could see the Potomac River, bubbling over rocks. They passed Georgetown University, with its spires and domes. Finally, rounding the bend onto F Street, they could see it.

In 1958, President Dwight D. Eisenhower, worried that the world thought of the United States as greedy and domineering, came up with the idea for a national cultural center.

Architect Edward Durrell Stone, a professor of architectural history at Yale University, designed the modern building with a nod to the ancient Greeks and Romans. The 100-foot-high structure was built with 3,700 tons of Italian marble lining the interior and exterior walls—a million-dollar gift from that country's government. A portico was held up by sixty-six monoliths. A terrace provided a commanding view of the river.

Lisa's parents parked, and the family rode the escalator up into opulence.

They walked into the opulent Grand Foyer of the Kennedy Center for the Performing Arts, their feet sinking a little into the red carpet, as eighteen crystal chandeliers glimmered in the morning light above them. Outside the giant panes of glass, they could see the trees of Roo-sevelt Island Park, shining like emeralds laid at their feet. Lisa stole a quick glimpse of her reflection in the towers of mirrors lining the foyer's walls—an American Versailles.

As her family found their seats in the building's largest auditorium, she joined her classmates in a tide of purple caps and gowns.

At 9:30 A.M., the Woodward Orchestra began the familiar strains of Elgar's "Pomp and Circumstance" while the graduates snaked in lines to their plush, red-velvet seats. Among them was one of New Mexico senator Pete Domenici's daughters. The dignitaries onstage included the morning's guest speaker, Senate Majority Leader Howard Baker, a Re-publican from Tennessee. Beside him were Carole Wallace, president of the Montgomery County School Board, and Principal Anita Willens.

Dangling above them in the 2,442-seat auditorium were eleven Had-elands crystal chandeliers, gifts from the government of Norway. The box where the president and first lady would sit when attending events was in the center of the first balcony. Six months prior to this ceremony, the Kennedy Center and President Jimmy Carter honored the great U.S. composer and conductor Leonard Bernstein, actor James Cagney, cho-reographer Agnes de Mille, stage actress Lynn Fontanne, and the late opera soprano Leontyne Price. On that very stage, the National Sym-phony Orchestra plays each season. In the previous year, the Kennedy Center had celebrated Aaron Copland's eightieth birthday and Isaac Stern's sixtieth.

And here she was, tiny Lisa Caputo. Born in the District and raised in Rockville, a town of money and high achievers in one of the wealthi-est public school systems in the country. She listened as Rabbi Joshua Haberman said a prayer. Then she tried to absorb what the senator was saying.

Senator Baker had become world-famous nine years earlier when he served as the ranking minority member on the Senate Select Commit-tee on Presidential Campaign Activities. The committee had questioned White House aides about the break-in by the Committee to Re-Elect the President at Democratic National Headquarters next door at the Watergate.

With his thick black glasses, he was memorable as the man who re-peatedly and famously asked, "What did the president know and when did he know it?" As the 1972 summer hearings wore on, Baker's presi-dential pendulum swung from innocent to most certainly guilty, and he helped to persuade President Richard Nixon to resign from office. It earned him the ire of his party but the respect of millions of Americans. In 1980, he had tossed his hat into the ring to run for president but then backed Ronald Reagan instead. And Lisa was sitting just a few feet from him.

After several songs, Principal Willens spoke. Then it was finally Lisa's turn. She wanted to say something—anything—that would inspire her classmates on to greatness. In the end, like nearly every valedictorian speech ever given, no one remembered a word.

Many of her classmates would go on to notable places like Harvard, MIT, and Duke.

When it was all over, after the tassels had been switched over and the caps had been tossed up, Lisa Caputo took her first steps into a world waiting for her to conquer it, launched from a national historic landmark into the sunniest of days in Washington.

4

★　★

The Victim

Colleen Shipman waited alone in a small, cold, interrogation room, her eyes still watering a little from the pepper spray that had been squirted into her car in an airport parking lot.

Why me? Why had this bizarre woman attacked me? she wondered.

At his desk, Detective Becton read Shipman's handwritten, matter-of-fact, two-and-a-half-page statement. Then he walked into the room where Shipman was waiting. She looked very meek and worried, and asked if she could call her commanding officer at Patrick Air Force Base, where she was stationed as a captain. She helped with military rocket launches at Cape Canaveral Air Force Station, a few miles north of the base.

Shipman began telling Becton all she knew at that point, the story of this crazy woman following her and attacking her. She talked about first noticing the woman on the shuttle bus, her hood from her tan trench coat pulled low and her jeans rolled up into two matching cuffs. Her black shoes. A small woman with short dark hair and big red glasses. She talked about how concerned she was with this woman's presence in the parking lot. The woman had followed Shipman to her car after they rode the airport shuttle bus to the blue parking lot, tried to get into her car, slapped at her window, asked her for a ride back to the terminal, asked to use her phone, and then pepper-sprayed her when Shipman refused to let this woman in her car.

Knowing they had the contents from the duffel bag and garbage can, along with this very credible witness, Becton began to think that this case was going to be a slam-dunk to close. Not only was she articulate, but she was also one of the most observant and descriptive victims Becton had interviewed in a long time. Becton became more impressed with her as he listened, thinking she had a really good head on her shoulders.

"Have you ever seen this woman before?"

"No, never."

"Are you an astronaut?" Becton asked her.

"Well, no, I'm not an astronaut," she told him, touching a gold space shuttle charm dangling from her necklace. "But I wear a shuttle necklace because my boyfriend bought it for me. *He's* an astronaut."

"So, you're dating someone?" he asked, already knowing she was. She said yes.

"Does the name Lisa Nowak mean anything to you?"

"I think that's the name of my boyfriend's friend. She and Bill dated for a while, but Bill called it off after he and I started seeing each other in November."

"Have you ever seen Lisa Nowak?"

"Um, just from the picture of the shuttle crew."

Then Becton saw that the light was coming on as Shipman started to realize why he was asking these questions about Nowak.

"Who attacked me?"

Becton hesitated a moment. He wanted to make sure to study her when he told her.

"Lisa Nowak."

Colleen Shipman's face went white, and her eyes had the look of a rabbit with a paw caught in the teeth of a trap—utter terror. She immediately came to the conclusion that Lisa Nowak had planned to gravely hurt her, if not kill her.

Shipman asked if she could call her boyfriend. Becton let her make the call on the office landline. She woke up Billy at his apartment in Houston, the one she had left behind just a few hours earlier. She began to tell him what happened.

"Baby, was your friend Lisa Nowak? I think she's here. I think she just attacked me."

She dissolved into tears as she recounted the story of this crazy woman pepper-spraying her. She handed Becton the phone. On the other end was a groggy male voice, asking if Colleen was okay.

As Shipman continued to cry softly, Becton stepped out to his desk to talk privately with Navy Commander William Oefelein.

The astronaut confirmed that he was dating Shipman and that he and Lisa were "good friends." He was reluctant to admit any kind of intimate relationship with Nowak. After a few minutes of chatting, he slowly said that, well, yes, there was more to it, but it was mainly a working relationship.

After the call, Becton had questions. He needed to figure out how Lisa knew when Colleen's flight was getting in.

Did Bill set all this up so he and Lisa could be together? Or was Nowak stalking Shipman to kill her so she and Bill could be together? But how did Nowak get to Orlando?

Becton didn't really believe Bill because he had been evasive about his relationship with Nowak.

The detective went back to his desk, where a patrol officer handed him Nowak's biography, printed out from NASA's website. For Becton, it was like reading a "Who's Who" of the military and NASA. The Naval Academy, test pilot, the astronaut corps . . . it was the stuff of legends. And of one very smart woman.

He called Oefelein again. How would Nowak know when Shipman's flight was getting in? Did he tell her?

"No. I had told her Colleen was coming."

"Does she have access to your computer or your emails?"

"Lisa had a key to my apartment."

Becton hung up with Oefelein. He glanced at the clock—6:30 A.M. He needed to figure out how to keep Nowak from outsmarting him.

I'm just a street cop having to interview a damn rocket scientist and combat jet pilot, he thought. *How am I going to do this?*

Becton had graduated from the police academy at Seminole Community College in the late 1980s and had occasionally returned to lecture new recruits. Despite his epilepsy, he had fought and won the right to join the police force in upscale and quiet Winter Park before transferring to Orlando. But his career was nothing compared to Annapolis and Johnson Space Center. He knew reverse psychology wouldn't work.

This interview is going to have to be straight up, on the table, he thought. *I will have to keep control of the interview without violating any rules, policies, and laws.*

Becton paused a few minutes to pray and read a daily Bible verse, knowing that he needed guidance and the strength of his faith to get through this. He also made a phone call to Bridgette because he knew there would not be time to check in with her every two or three hours like they usually did. This case was just getting started.

5

★ ★

Anchors Aweigh

Beyond the Ivy League schools, there are only four places in America that are more elite and more demanding: the United States military academies. And Lisa Marie Caputo had been accepted into the United States Naval Academy (USNA) at Annapolis, Maryland.

Not only did she have to be at the top of her class to get in, she also had to be physically fit and a good citizen—her participation on her high school field hockey team, and her role on student council helped meet those criteria.

More than fifty U.S. astronauts have graduated from the Naval Academy, including moonwalkers Alan Shephard and Charles Duke, along with Apollo 13 commander Jim Lovell. Former president Jimmy Carter is a graduate, as is Nobel laureate Albert Abraham Michelson, the first American to receive the Nobel Prize in science for his measurement of the speed of light. In addition, more than seventy Medal of Honor recipients are alumni of the Naval Academy, including former U.S. vice-presidential candidate and Vietnam War POW James Stockdale.

According to the Naval Academy, about 2,000 candidates are fully qualified each year for admission. Of that number, about 1,500 are offered appointments, and approximately 1,200 actually become midshipmen—and some of the most focused, disciplined students in the world.

Not all enter as Nowak had; some entered as enlisted servicemen and women, while others transfer in from colleges or prep schools.

Lisa began her application process just four years after Congress began allowing women into all four military academies. Requirements for women included being a U.S. citizen, having good moral character, being unmarried and not pregnant, and having no dependents. The academy also wanted to know her rank in high school, along with her college test scores.

Another hurdle to jump: she also had to obtain a nomination from her congressman, senator, or the U.S. vice president.

Because of her outstanding academic record, athleticism, and leadership capabilities, she was most likely one of those midshipmen lucky enough to receive an "official candidate for admission" letter in July 1980, the summer before her senior year of high school began.

In the early morning of July 7, 1981, Lisa headed to Alumni Hall, along with the other new midshipmen who had been appointed and accepted.

"Good morning, Shipmate. Welcome to the Naval Academy," she was greeted as she walked in.

For the next few hours, she learned everything a midshipman needs to know to start at the academy, including how to salute and how to properly wear her white cap with a black stripe—denoting a plebe. She received books and gear and stuffed them all into a large, white laundry bag that she marked with her name. She gave up her civilian clothes and received white shoes, white pants, a white jacket, and white belt.

When she reemerged into the hot sun, her hair was cut short to her ears—her male counterparts all had short buzz cuts—and she and the other plebes marched in formation to take their seats in Tecumseh Court, where they were all sworn into the U.S. Navy.

"Would the class of 1985 please rise and raise your right hand," Academy Superintendent William Porter Lawrence ordered. "Having been appointed midshipmen in the United States Navy, do you solemnly swear that you will support and defend the Constitution of the United States against all enemies, foreign and domestic, that you will bear true faith and allegiance to the same, that you take this obligation freely, without any mental reservation or purpose of evasion, and that you will well and faithfully discharge the duties of the office of which you are about to enter—so help you God."

In unison, the midshipmen shouted, "I DO!"

Also sitting in the crowd that day was another young plebe, Richard T. Nowak of South Burlington, Vermont. In April 1988, the pair would say, "I do," again at the Naval Academy—this time to each other, in their wedding ceremony.

Midshipmen are referred to differently from regular college students:

The equivalent of a freshman is called a Plebe or midshipman
fourth class;
the equivalent of a sophomore is called a Youngster or midshipman third class;
the equivalent of a junior is called a midshipman second class;
and the equivalent of a senior is called a Firstie or midshipman
first class.

After visiting with family, the plebes made their way to their rooms.
Lisa became the butt of jokes for an event that happened following the
swearing in—she got lost for an hour trying to find her room in the
enormous Bancroft Hall, which houses all 4,500 midshipmen.

"Even to this day, Lisa has retained her incredible sense of direction,"
her senior yearbook entry, written by fellow classmates, sarcastically
reads.

Lisa then set out on a Naval Academy tradition, dating back to the
school's earliest days after it was founded in 1845 by Secretary of the
Navy George Bancroft on ten acres of Fort Severn, overlooking the
Severn River.

It's known as Plebe Summer and is considered by some to be pure
hell. It's seven weeks of grueling exercise beginning at dawn and ending
"long after sunset, wondering how you will make it through the next
day," the academy's website reads. Days are filled with military drills,
military exercises, and the extremely structured rigors of military life.

If it were difficult for the men in her class, it was even more so for
Lisa and the few other women there. Many instructors were openly hostile to the presence of females in that storied male bastion.

For seven weeks that summer, Lisa and her fellow plebes learned to
carry one another—literally and figuratively—as they were separated
from all and everyone they had known before.

Plebes learned to sail on the Severn River, to understand the time-
less tradition of being powered by only the wind and the current. They
learned to work as a group as they paddled rubber rafts into the current. And they came to understand their own fortitude as they crawled
through muddy trenches, under barbed wire, looking like World War
I doughboys. On the firing range, many got their first opportunity to
shoot a gun.

They stood watch day and night, and marched in formation to

the dining room. Part of learning discipline was to keep their rooms "squared away" at all times—ready for inspection at any moment. Those whose rooms were found to be less than spotless received demerits.

"You learn self-discipline. You learn to organize your time and decide which things are most important. You reach top physical condition. You develop your ability to think clearly under stress and to react quickly when the unexpected comes your way," the Naval Academy website states. "Any officer who has stood the watch on the bridge of a ship in a storm or landed a jet on the deck of an aircraft carrier at night can tell you the importance of these qualities."

But most importantly to Naval Academy officials, the plebes learned about the "high standards of honor, character and conduct." They learned, as more than 110,000 others before them, the importance of seeing a task through to the end. "Don't Give Up the Ship," the dying words of an early U.S. Navy captain, is a phrase that became embedded into their conscience.

As Lisa began the fall semester, her studies included engineering, science, math, humanities, and social science. Officials wanted each midshipman to qualify for any career field in the Navy or Marine Corps. As Lisa dreamed of flying into space, she concentrated on aeronautical and astronautical engineering.

Her typical day began that fall at 5:30 A.M. with an optional personal workout. Reveille was blown on the trumpet at 6:30 A.M., with everyone expected out of bed immediately. Plebes got "special instructions" until 7:00 A.M. (basically, harassment by older midshipmen), then lined up to march to breakfast. All 4,500 midshipmen eat every meal together in the vast King Hall. After breakfast, midshipmen attended four classes of less than an hour each; then it was time for lunch and, afterward, training for a half hour. Two more classes were held after that, then time for extracurricular activities, with drills and parades held twice weekly until 6:00 P.M. Everyone cleaned up for dinner, ate, and then studied until taps was played at midnight.

Midshipmen were not left to float in a vast sea of other bewildered students. They became part of one of thirty "companies" in Bancroft Hall, made up of about 150 students each. Of course, roommates and people on the same floor grew close in units, but those companies "eat, sleep, study, drill, play and compete as teams," working toward the title "Color Company"—the best of the entire academy. The title alone could be the reward, but the color company also has the privilege of representing the Naval Academy at official functions, including presidential inaugurations.

Company members learned to trust and rely on each other in what is known as "small-unit cohesion," an invaluable way of living in the military and surviving during wartime. Book learning is accompanied by the practical experiences learned in the units.

Free time, what little there is of it, was closely guarded by the Naval Academy. Plebes could go into town from noon until midnight on Saturdays only and could spend time within the academy complex—known as The Yard—on Sundays. Many spent time with a "host family" in the town, getting a taste of the home and family life many missed, and forged friendships that would last a lifetime. But midshipmen could not leave at all if they had duty or were struggling with academics, conduct, or military performance.

In the spring of her plebe year, Lisa and her classmates chose their majors—a full year and half before most civilian college students are forced to pick a course of study.

The plebes endured a year of what officials call training, some might say is harassment, and others would label hazing. But some of it is meant in jest. According to the U.S. Naval Academy parents' website, plebes are required to bet that Navy will win each football game. If that happens, they gain a few privileges, like watching television, playing a movie in their company's ward room, or listening to music. If Navy loses, though, the plebes can lose their butts for one week, forcing them to ask—"always at the top of their lungs"—to borrow them back for everything from using the toilet to sitting down for dinner. Or, they can lose the ability to turn right, forcing them to make three left turns in order to go right.

During the week in May when first-year students graduated and were commissioned as officers into the U.S. Navy, the plebes participated in a tradition dating back decades. They climbed on each other to reach the top of an obelisk. It is a monument to Commander William Lewis Herndon, who died trying to save the SS *Central America* during a hurricane off Cape Hatteras, South Carolina, in 1857. The obelisk was covered in a thick layer of grease, and the plebes tried to reach the top of the two-story-tall monument to retrieve a white cup and replace it with an upperclassman's hat. Lisa's class had the distinction of having one of the slowest times on record of accomplishing this task: 3 hours, 12 minutes, and 23 seconds. The fastest time was one minute, 30 seconds, held by the class of 1972, when the monument was not greased.

Finally, the group participated in Sea Trials, a one-day "final exam" of all the military maneuvers they had learned throughout the previous year, including: carrying a heavy ammunition box across Lejeune Pool and back without getting the box wet; assembling an object at the bottom of a pool; escaping from a POW camp; running an endurance course through the toughest conditions; enduring interrogations in a smoke-filled tent before a fake bomb "explodes, 'killing' their comrades"; and rope-crawling across a river.

Those who survived their plebe year at Annapolis were rewarded with what's known as a 3c cruise. These could be on ships as far away as the Western Pacific, the Mediterranean, or in the South Atlantic off the coast of South America.

But Lisa received the least-desired duty: what was known as the Yard Patrol cruise, which took participants on one of about a dozen 110-foot boats along the U.S. east coast, stopping at ports like Newport, Rhode Island, New York City, and Boston. It gave midshipmen their first salty taste of life aboard a boat, and they learned to perform every operation on board—from kitchen patrol duty to captaining the ship. They also learned to chart courses and navigate, taking what they had studied in class out onto the high seas.

Lisa returned to Annapolis after that cruise, and, upon spotting the dome of the USNA Chapel, she officially became a "youngster," or midshipman third class—and a Plebe No More. She could sit on the benches in The Yard, stay up later than 11:00 P.M. (if she weren't completely exhausted), and watch television (if she had the time).

With her family just an hour away in Rockville, Lisa could also easily head home for weekends after the discipline of her first year.

Her Youngster Year involved deeper studies as she concentrated more and more on aerospace engineering. It was a department at the academy designed to, among other things, groom future astronauts, as evidenced by the more than four dozen of them who have walked across The Yard as students. Some of the courses she took included astrodynamics, rocket propulsion, space environment, spacecraft power and communications, spacecraft vehicle design, wind tunnels, and flight propulsion.

In the spring of their second year, Lisa and her classmates partici-pated in the ring dance—a formal event at which they received their coveted class rings. It is a piece of jewelry many will wear proudly for the rest of their lives. Each year, a different crest is designed, and receiv-ing the ring is on par with slipping on a wedding band. The rings are blessed by Navy chaplains, who dip them into a bowl of water gath-ered from the world's seven seas. After dinner and dancing, the night is capped off with a fireworks show.

At the end of the year, the Youngsters gathered on Hospital Point for their traditional luau. They designed their own T-shirts and sported leis and shorts.

During the summers of her third and second years, she took advan-tage of military flights to travel to Hawaii, France, and Germany. She also spent time in Pensacola for flight training.

As a second-class "Mid," Lisa gained more freedom but endured what is generally described as the toughest academic year at the acad-emy, with coursework dedicated to her major study area. She and her classmates were also in charge of the plebes, experiencing leadership as a naval upperclassman and subjecting the freshmen to the same rigors they had endured.

Before moving on, Lisa passed two more grueling physical activi-ties—the Tower Jump and the 40-Year Swim. The latter involves swim-ming half a mile in forty minutes. The catch—she and her classmates swam fully clothed.

Before the year was out, Lisa and her classmates underwent a battery of physical exams, measuring their body-fat index, vision, height, and arm-reach numbers. The results were all combined to see for which ca-reers they qualified. Without the proper vision and arm reach, aviation was out of the question.

And finally, they are "Firsties"—midshipmen first class. Lisa chose a second tour of Pensacola Naval Air Station and a trip on board the USS *Lexington,* a World War II–era aircraft carrier, for her first-year cruise in the summer before her final year at the academy. She acted as a junior officer, standing bridge watch and, at times, controlling the ship.

After a vacation in Europe, she was back in Annapolis for her final year. On her uniform, she wore a star and at least one horizontal stripe on her sleeves and shoulder boards.

In May 1985, Lisa celebrated Commissioning Week with her 1,000 classmates. They attended the Superintendent's Party and the Graduation Ball. On May 22, they donned their dress white uniforms and marched out onto Tecumseh Court one final time as midshipmen, with President Ronald Reagan looking on.

"Sometimes it's hard to find the words to express my heartfelt gratitude for those who serve on active duty and in the reserves," Reagan told them. "But it isn't difficult to find the words to explain why they do what they do. It only takes one word—patriotism. And as Commander in Chief, I am overwhelmed at times by their dedication and courage."

And then it was time for that most iconic moment at the Naval Academy. After being commissioned as officers in the United States Navy or Marines, the 1,000 graduates tossed up their white Naval Academy caps, a tradition dating back to 1912, when the midshipmen were first issued officers' hats following the ceremony.

In the weeks before the ceremony, the Firsties received their marching orders. Those at the head of the class got the first pick of where they wanted to go. Lisa's orders: temporary duty at Johnson Space Center in Houston, where she would work with astronauts and Mission Control for six months.

At Johnson Space Center, Lisa was impressed with the dedication of everyone in the space program as she worked in Capsule Communication—called CAPCOM. Some days, she got to work in Mission Control. She described it as "an exclusive job." She also was able to work with the robotics department. She helped to design training and rewrite instructions.

"I had the opportunity to come and work here at NASA for six months for temporary duty before going off to flight school and I got to meet everybody in the program here—not just the astronauts, but everybody that works in all the different jobs," Lisa said in an interview with NASA. "It seemed really exciting seeing it up close. I thought if there was a chance to be able to come here that I would love to do that. . . . [W]hat impressed me was the whole idea that everybody was so into what they were doing and excited that each of their parts was so important. I think that is probably the biggest thing that excited me was that all of the people were so into the mission and knew they were a big part of it."

In January 1986, Lisa Caputo and Richard Nowak both reported as student pilots to the United States Navy Flight School in Pensacola. It was here that the pair began to date.

Flight training for Naval officers is almost as old as aviation itself. The Wright brothers made their historic first sustained flight at Kitty Hawk, North Carolina, in 1903, and the U.S. Navy began training pilots in Pensacola eleven years later.

Until the school was moved to Newport, Rhode Island, in 2008, officer candidates could often be seen running through the streets of Pensacola in blue shorts and white T-shirts. Those who pounded the sidewalks include aviation pioneer and legend Neil Armstrong, and future U.S. senator John McCain. Often screaming through the air above them were the Navy's precision air show team, the Blue Angels.

Of course, even more legendary were the drill instructors, getting in candidates' faces, barking orders, demanding more than the candidates thought they were capable of giving.

"Every candidate I've ever talked to always remembers their drill instructor because they are such a dominating presence, a larger-than-life presence," base historian Hill Goodspeed said in a 2008 interview.

And when the drill instructor was displeased, there was always the march down the seaplane ramp into Pensacola Bay, M-1 rifles held above their heads as they treaded water.

The recruits barked in unison to answer questions or announce their plans.

"SIR! FORTY-FIVE! HIGHY MOTIVATED!! TRULY DEDICATED!! AVIA-TION!! OFFICER!! CANDIDATES!! MA-ARCH INTO CHOW!! AYE AYE, SIR!!!"

There were the infamous, grueling obstacle courses and ten-mile-long "rifle runs," carrying a 9½-pound M-1 rifle in 90-degree heat and humidity of an equal number. The candidates ran it in eighty minutes, pounding the road in step and singing "Jodies," songs sung in a call-and-response fashion. According to Roger Morris, who graduated in 1979, "the rifle kept you from swinging your arms, the singing kept you from breathing naturally—it was an experience I was quite happy not to have to repeat."

The officer candidates also endured survival school at Eglin Air Force Base, near DeFuniak Springs, Florida, slogging through a swamp to learn navigation and how to avoid the local fauna, including alligators and the dreaded cottonmouth moccasin, which sometimes seem to pursue a person out of spite.

Both Richard and Lisa earned their golden wings as Naval Flight Officers in June 1987.

While Richard was sent to serve on an aircraft carrier, Lisa remained in Pensacola and became a trainee at Electronic Warfare School at Corry Station. She also received initial training on the A-7, an attack aircraft, at Naval Air Station Lemoore, California. She learned how to detect, interpret, identify, and locate electronic signals from weapons' radar emissions. She also determined effective defensive maneuvers and operated systems that produced high-powered electronic signals, used to deceive and jam enemy electronic sensors.

One of her roles was to help defend her ship against antiship cruise missiles.

At this point, she and Richard had been dating a year and a half. With him away on an aircraft carrier, their communications were few and far between. She later told a psychologist that he rarely wrote and only called when he was in port. She finally asked him what his long-term intentions were for the relationship, and that's when they decided to marry.

All the arrangements were made months ahead of time—they would return to where it all began and stand under the circular cupola of the United States Naval Academy Chapel to recite their vows in a Catholic ceremony. The archdiocese required the couple to go through either an "Engaged Encounter Weekend" or a full day of premarital counseling. They also had to fill out a prenuptial questionnaire. Among the questions:

> Do you intend to enter a permanent, indissoluble marriage and be faithful to your spouse for life?
> Do you intend to enter a marriage whose purposes are the mutual love and support of the spouse?

The sun rose on a frigid, foggy Saturday morning on April 9, 1988, with a little rain falling and winds whipping up to thirty miles per hour.

The ceremony was set for the chapel, but it is not a small, quaint building, as its name implies. Instead, it is a large structure, among the most iconic buildings and the focal point of the academy. Its copper dome had turned green since it was dedicated in 1908, although a recent cleaning and repair has it the color of a penny again. The building is shaped in a cross, with a rounded cupola over the altar and a window with a web design allowed light to fall onto the altar. A U-shaped balcony framed the second level at the back of the chapel.

Richard and his groomsmen were ordered to arrive first at the church and wait with Navy Chaplain Commander Frank Mintjal in a room on the lower level of the main chapel, near where John Paul Jones's sarcophagus is kept. Ushers were asked to wait at the door to seat guests.

Lisa and her bridesmaids arrived a few minutes later and waited in a bridal room at the back of the church. The couple hired a cantor through the Naval Academy music department to sing.

At the appointed hour, Lisa Marie Caputo walked into the chapel and down the main aisle with her father, as her mother waited in a front row pew a few feet from Richard, who was decked out in his dress blues.

She and Richard climbed up a few marble steps to the altar and Chaplain Mintjal greeted their guests and prayed over the couple and ceremony. After a scripture reading, Lisa and Richard exchanged vows and rings and the chaplain blessed them. And then he proclaimed them man and wife, and Richard and Lisa Nowak walked down the aisle as a married couple.

After their wedding and honeymoon, she became a member of Pilot Electronic Warfare Aggressor Squadron 34 at Point Magu, just west of Los Angeles, California. According to the Navy, Point Magu supports the fleet electronic warfare community. Point Magu also supports fleet training and tactics development, particularly large-scale, major exercises, including fleet battle experiments in the Pacific Ocean. They test multiple types of missiles and include nearly a dozen different types of aircraft.

While there, Lisa flew an EA-7L, an attack jet with carrier and subsonic warfare capabilities, and an ERA-3B, known as the "Skywarrior," a strategic nuclear bomber. She worked supporting the fleet in small- and large-scale exercises with jamming and missile profiles. And she was very good at her job: while assigned to the squadron, she qualified as mission commander and electronic warfare lead.

Between 1990 and 1992 she was a graduate student at the United States Navy Postgrad School in Monterey, California, from which she received a master's degree in aeronautical engineering. During grad school, she met another ambitious pilot who would later become a crew member on her space shuttle flight—Mark Kelly.

Her next stop was Patuxent River, Maryland—colloquially known as NAS Pax River—where she became an engineer for the Systems Engineering Test Directorate. The air station is on the Chesapeake Bay near the mouth of the Patuxent River—about a two-hour drive from her parents' home in Rockville.

While there, she was selected for both aerospace engineering duty and Test Pilot School. She told *Catholic Standard Magazine* that she had applied six times for the coveted test pilot position. Then she was named the aircraft systems project officer for the Strike Aircraft Test Squadron, also in Patuxent.

Her son was born in February 1992, and she remained at Patuxent River working as an aircraft systems project officer at the Air Combat Environment Test and Evaluation Facility and at Strike Aircraft Test Squadron, flying the F/A-18 and EA-6B jets.

Nowak was then assigned to the Naval Air Systems (NAS) Command, working on acquisitions of new systems for naval aircraft, also at Patuxent River. The NAS Command provides support for aircraft and airborne weapon systems, including research, design, development and systems engineering, acquisition, test and evaluation, training facilities and equipment, repair and modification, and in-service engineering and logistics support.

It was here that she met a newly selected test pilot named William Oefelein, who arrived in January 1995.

Around this time, she and about half of the officers she worked with at NAS Pax River applied to become astronauts. More than 2,500 people sent in applications. Of that, 120 people were chosen to travel to Houston for a week of interviews and rigorous physical exams. In April 1996, Nowak was one of thirty-five people selected to become an astronaut, moving to Houston that year with her husband and son. Her fellow test pilot at NAS Pax River, Mark Kelly, was also chosen that year, along with his twin brother, Scott.

In addition to working in CAPCOM, Lisa Nowak trained as a flight crew medical officer, receiving many hours of medical training within eighteen months before the mission.

According to an article in the July 2017 *Harvard Business Review*, "before launch, all astronauts are trained to use the medical assets that will be on board. There is not always a doctor on board, so some astronauts also undergo 40 hours of paramedic-level training to qualify as a crew medical officer. They become familiar with a checklist of foreseeable medical problems and emergency responses, such as a crewmate having difficulty breathing. They learn how to perform a periodic basic physical examination and how to handle the most common medical problems, such as motion sickness, skin irritation, and back pain. After training, they'll know they can either call their flight surgeon, who might direct them to a drug in the medical kit, or use a procedure to relieve the pain."

NASA's "Space Medicine Exploration Medical Condition List" contains at least five dozen conditions that crew medical officers must treat while on board the spacecraft. They include things as simple as a rash and burns to far more serious issues like appendicitis, tooth extractions, and "surgical treatments."

Astronaut Clayton Anderson lived on board the International Space Station (ISS) for 152 days in 2007 and again for twelve days in 2010. He trained as a flight crew medical officer.

"As the crew medical officer, I received special—and additional—training for handling several key scenarios," Clayton said in an interview for gizmodo.com. "For example, I would be counted on to collect blood samples, perform urinary catheters (has actually happened in space!), execute defibrillation, intubations, the administering of sutures, basic dental (pull a tooth) and eye care (search for foreign objects)."

Within about a year after the Nowaks' arrival in Houston, she and Richard began "growing more distant, physically and emotionally. Despite this, Lisa Nowak was determined to make the marriage last, in part due to her Catholic upbringing, but also [out of] a desire not to hurt her son and cause embarrassment or emotional pain for her family of origin," she told a psychiatrist in 2007.

6

★　★

The Chess Match

Detective Becton took out the tape recorder and microphone from his desk. He knew he would have to record this conversation and get it right, but he didn't want to intimidate Nowak or shut her up by shoving a microphone in her face. He had conducted other interviews without a microphone, and they came out fine. He started to walk to where she was being held, the tape recorder in his hand. But he put the mic back in the desk drawer—a decision he later deeply regretted.

It was nearly 7:00 A.M. She had been in custody for about three hours. She sat behind a cheap Formica desk that was bolted to the wall. She was so small, she looked almost childlike sitting in the black plastic chair. But Becton thought she also looked hard. It was as though her heart and soul had been taken from her, and all that remained was this husk of a woman.

Becton introduced himself to her, but his phone began to ring with his home number flashing on the screen. He excused himself and stepped into the hallway. Becton's youngest daughter, ChristianAnn, was crying because he was not there to take her to school. Becton empathized with his wife—he knew she had her hands full getting their three children ready for school without him there to help, as he usually was. And their youngest—Becton's daughter from his first marriage—could be difficult sometimes. Becton calmed her down and asked her to put Bridgette back on the phone.

"I love you, honey. Thanks so much for everything you do," he told her.

"Love you, too."

Then he stepped back into the interview room and sat down across from the astronaut.

"You've been placed under arrest for what's happened in the parking lot. You're not willing to talk to me; just say so now and we'll deal with what we've got and let everybody interpret what they interpret out of what we've got. If you would like to tell me your side, then we'll go ahead, and we'll continue and you get a chance to say what it is you want to say."

"How can that help me?"

"Well, what you say can change what you're charged with, okay. Um, right now, we're looking at a possible life felony for carjacking."

"You think I was trying to steal a car?"

Becton knew he needed to get through the Miranda rights. If she wanted an attorney, by law, he would have to stop the interview. He also knew that if he were under arrest, he wouldn't say jack diddly to any police officer.

"Are we at that point where I can let you know what your constitutional rights are?"

"Give them to me, yeah."

"All right. First, you have the right to remain silent. Do you understand?"

"Yes."

"Anything you say, and let me, before I go any further, did you ever watch *Cops*?"

His mind flashed to the Saturday-night FOX show he and his stepsons loved to watch. They always picked the stupidest, meanest, most screwed-up people to put on TV.

"No."

"Okay, um. Let me continue this, anything you say can be used against you in court, do you understand?"

She mumbled something Becton thought was yes. He deviated from the standard script, hoping to help her understand what he was saying.

"You have the right to talk to your lawyer before and during questioning, do you understand?"

Again she mumbled.

"Yes?" he asked.

"Yes."

"And you understand that any, whatever you tell me is gonna be used in court?"

He noticed something that he had seen in other suspects. She looked like she was on something—some kind of drug, maybe an amphetamine. She was not showing any emotion—no sadness, no fright, no anguish, no worry. Nothing. He doubted in that moment that she would talk to him.

"Do you advise that I need a lawyer?"

Becton's head tilted a little. He was shocked.

Is this some kind of game? An astronaut who has flown on the shuttle, who has trained on numerous fighter airplanes, this brilliant woman who has physically attacked another woman is asking him if she needs an attorney? Is she kidding? Is she doing what the military taught her to do if she were ever shot down behind enemy lines?

"Well, let me finish reading this, and then maybe I can answer your questions, okay? If you cannot afford a lawyer and want one, one will be provided for you before questioning without charge."

He had finished informing her of all her rights to silence and to an attorney.

"Do you understand?"

"Ummmmm."

She looked like she might want to talk to him, but only to see how much he knew. And he knew he had to share something with her to gain her trust, but he had to be careful not to reveal too much. How could he get her to trust him enough to open up? She was stone-cold and seemed to be analyzing him and his questions.

"Has anyone threatened you or promised you anything to get you to talk to me?"

She mumbled again. "No."

"You do have the right to talk to a lawyer before and during questioning. You can stop questioning any time you want. However, we go back to it's a lopsided situation, okay. There's obviously a reason why you're here tonight. Something has happened to you, and if you're not gonna tell me what . . . if you're not gonna tell me why you're here tonight, then you're kind of on your own, okay? Again, I cannot make you any promises about anything. I'm not authorized to do that. But I can step up to the plate and speak up for you."

"About what?"

"Well, you know, if there's a logical explanation to all this, it's very easy when it comes time for me to discuss with the state attorney's office what's going on to say, 'Well, you know, have you read her bio?'"

"It doesn't mean anything." She was speaking in almost a whisper, as though the astronaut in her barely even existed anymore.

"Actually, it does. This is pretty impressive. For you to be here, something serious has happened to you. I mean, you know, when you look at the fact that you have 1,500 flight hours in more than thirty different aircrafts, that's impressive. You're a Navy test pilot. You went through Navy test pilot school. Is that where you got picked to be an astronaut?"

"Yes."

"You were selected for the Navy Test Pilot School, that's impressive. Okay—so something had to happen for tonight to go down. And if you need help, you know, let's face it . . . everybody needs help at some point in their life, and you can't always do everything by yourself."

She thought about that for a minute. It seemed to register with her that she was in a very bad situation and she needed someone's help. Although Becton had meant professional psychiatric help, she suddenly seemed to think he was the one who could get her out of this.

"I really needed to talk to her," she said. "I mean, I just wanted to sit and talk."

Becton knew he had just broken through her wall. He had to move gently to keep her going.

"Do you even know her?"

"I've never met her before—in person."

"So what would you talk to her, what, what would you say to her?"

"I would . . . ," her voice trailed off. "A matter I think we had in common."

"Okay, do you mind telling me what that matter is?"

"You said you've talked to her?"

"I have talked to her."

"Then I assume you already know. I don't want to bring other people into it, if it's not necessary."

Becton saw that she was trying to stay a step ahead of him. It would be the beginning of what he saw as an hours-long chess match. He would make a move; she would countermove. But he would eventually see that he was a pawn to her queen.

"She tells me she doesn't know you. She's never met you."

"We've never met. That's true," Nowak said.

"Let me do this . . . and I kind of see where you're going with this . . . you're trying to protect somebody from a lot of, um, undue stress, can we say? So let's start with, um, we won't use real names, how's that?" he asked her.

"Can you tell me what she talked to you about? That would help," Nowak said.

"Well, once you've told me your side, then I can go ahead and verify that. But there's no sense in me giving you everything she said, 'cause then, why even talk to you? I need to hear your side. I tell you what, let me ask this, do you mind if I refer to him with a different name?"

"You can say whatever you want," she said.

"Okay. Well, we'll say Tim. The two of you somehow, I know what her connection with him is. I don't know what your connection with him is. Can you tell me what is?"

Pawn to king four.

"Well, maybe that's what we needed to talk about," Nowak said, chuckling a little. It was her first real show of any emotion.

He knew she ran the robotic arm on the space shuttle, a task in which she would have to think in advance of her next move. Of her next ten moves. And he knew that's what she was doing now.

"It's like I've already been overcome by her wishes," she said. He noted that she felt like she wasn't in control anymore. "I, I, I believe it's not in her wishes to do that. Does it make sense?"

"Yes it does make sense. And right now, I'm not really concerned with her wishes. I'm more concerned with yours, okay? She's had her chance to tell me what's going on. I'm giving you that chance. I'm trying to find out—"

"Did you tell her who I was?"

"Yes, I mentioned your name. She says she's never met you."

"But she knows me?"

"She knew of you."

"Okay."

"But she's never met you. Now, I see you're a captain in the Navy. So you're not stupid; you're extremely intelligent, especially to be where you are today."

"Wait, I would like to go forward."

"Okay, I, I can help you do that once I know what's going on in your life."

Checkmate. Maybe.

7

★ ★

Columbia Disaster

On the chilly morning of February 1, 2003, at her Houston home, Lisa was watching NASA television as *Columbia* reentered Earth's atmosphere. The channel shows nearly everything before, during, and after flights and is on standard cable on Florida's Space Coast, in Houston, and elsewhere throughout the United States.

Lisa's husband, Richard, normally worked in the communications center, the large room where dozens of technicians speak with and monitor the astronauts and the shuttle. According to *Texas Monthly* magazine, he had been called up to active duty in the U.S. Navy the year before as part of America's War on Terror, and Lisa was balancing raising three children, performing her duties at NASA, and running a household, with the help of a nanny.

That morning, Lisa was waiting to see the safe return of her colleagues. She had gone through astronaut training with three of them: Pilot William McCool, Dr. David Brown, and Dr. Laurel Clark. Also on board were Air Force Colonel and Space Shuttle Commander Rick Husband, Israeli Air Force Colonel Ilan Ramon, Air Force Lieutenant Colonel Michael Anderson, and Mission Specialist Kalpana Chawla.

McCool was *Columbia*'s pilot and a rookie astronaut. His excitement about being an astronaut and going into space was barely containable.

"One of the biggest challenges that I'm faced with oftentimes is being able to push that excitement onto the back burner and to stay operationally focused on the tasks at hand," he told *Florida Today* space reporter Chris Kridler. "And that's, I'm sure, a challenge of every first-time flyer."

Dr. Brown might have had the most unique pre-astronaut career—he had run off and joined the circus. He helped to put himself through school at the College of William & Mary by working at nearby Busch

Gardens as an acrobat, tumbler, stilt-walker, and unicyclist. And then he worked for a traveling circus before going to medical school.

One of Lisa's best friends, Dr. Laurel Clark, was one of the only women to whom she had ever been close. They had sons the same age, and their families often got together. Dr. Clark was on board as a mission specialist, as a member of two teams that conducted more than eighty experiments nonstop during the fifteen-day flight. Her medical expertise was extremely specialized. She had completed Navy undersea medical officer training at the Naval Undersea Medical Institute as well as diving medical officer training at the Naval Diving and Salvage Training Center. She had performed medical evacuations from submarines during an assignment in Scotland.

Space travel is inherently dangerous, but launch and reentry are the most dangerous parts of any mission. During launch, the main fuel tank or solid rocket boosters could explode, as they did with the space shuttle *Challenger* in 1986. During reentry, the shuttle drops through Earth's upper atmosphere, a superhot layer of atmospheric gases surrounding the planet, at speeds as high as Mach 24—18,414 miles per hour. Black tiles, made out of a material called reinforced carbon-carbon, cover the leading edges of the shuttle's wings and nose cap, and black ceramic tiles protect the shuttle's underbelly. These tiles are designed to withstand the superheated temperatures of reentry and protect the ship's thin skin, then cool down after landing.

But that's not what happened on that February morning.

During launch two weeks earlier, as *Florida Today* newspaper would report, a pound-and-a-half chunk of insulating foam on the external fuel tank hit the leading edge of *Columbia's* left wing, knocking a small hole in the wing's tiles. A few NASA engineers knew about the hole and had tried to alert their higher-ups, but NASA officials did not think it was a problem. Shuttles had been hit by insulating foam on previous missions without consequence.

At approximately 7:15 A.M. Texas time, as *Columbia* slid eastward above the Indian Ocean, Commander Husband began a 2½-minute burn of the shuttle's orbiter maneuvering systems to drop *Columbia* out of orbit. He then positioned the ship for its first encounter with Earth's atmosphere as some in the crew began dumping the remaining fuel from the maneuvering system. Other crew members were putting on and testing the pressurization of their flight suits and getting buckled into their seats.

Video shot by Astronaut McCool from inside the upper flight deck shows Commander Husband trying to finish up a drink as he and McCool went through their checklist of things to do during descent. Rehydration was important for astronauts returning to Earth's gravity. At one point, he bumped the shuttle's joystick, a common occurrence on any shuttle flight due to the tight quarters.

Downstairs on the mid-deck, astronauts Ramon, Anderson, and Brown were stowing away the last pieces of cargo and dumping the remaining fuel and debris. Back upstairs, Husband, McCool, Chawla (seated behind McCool), and Clark (seated behind Husband) began testing their suits. All those astronauts took their cumbersome gloves off but left their helmets on. Chawla—K.C., as everyone called her— took the camera from McCool and began videotaping from her vantage point. She aimed the camera at Clark, who happily waved. Then K.C. aimed the camera at herself and jauntily smiled. Outside the windows, they could see flashes of plasma and aft jets firing as the ship began to enter the atmosphere.

Astronaut Mark Kelly explained in his autobiography *Gabby: A Story of Courage and Hope* that, "as the orbiter hits the atmosphere at such a high speed, an incredible amount of friction is generated. This friction strips the atmosphere of some electrons, creating plasma, which is a state of matter that has some properties of a gas but can flow like a liquid. It doesn't take long for the wings and fuselage to heat up to thousands of degrees, and you are literally flying in the middle of a giant fireball of plasma."

It's the high-altitude equivalent of quickly rubbing your hands together—eventually, it gets very hot from the friction.

"Ahhh, it's really neat—it's a bright orange yellow out over the nose, all around the nose," McCool said.

"Wait 'til you start seeing the swirl patterns," Husband answered as the ship started experiencing G-forces.

"This is amazing, it's really getting fairly bright out there," said the awestruck McCool.

"Yep. Yeah, you definitely don't want to be outside now," Husband said.

The ship sailed across the Pacific with no problems and approached the California coastline just before 8:00 A.M. Houston time. But then a sensor in *Columbia*'s left wing indicated a temperature rise. Thirty-nine seconds later, a sensor in the left wing became disconnected from the computer system, sounding an alarm in the cabin. Almost a minute later and midway over California, high-powered government cameras captured debris falling from *Columbia.* What the astronauts and NASA officials did not know was that, during reentry, superheated gases had flowed into the shuttle's wing and were now melting the ship's aluminum interior frame.

According to the *"Columbia" Crew Survival Investigation Report,* during the reentry period of any shuttle mission, the crew members typically experience heaviness, dizziness, and sometimes stomach awareness or mild nausea. Everything still seemed normal to the crew at this point, despite the "damage-induced aerodynamic changes to the orbiter." As usual, "the vehicle also began to experience the thermal effects of Earth's atmosphere. Shock waves due to the vehicle's hypersonic speed and the frictional effect of the atmosphere began to heat the orbiter's surface. Temperatures on the bottom and forward surfaces of the orbiter during entry vary by location on the ship, with the nose and leading edge of the wings experiencing temperatures greater than 2,800 degrees Fahrenheit (1,538 degrees Celsius)."

According to the report, *Columbia's* crew first knew there was a problem at 7:58:39 A.M. Houston time, when the first of four fault messages lit up on a monitor on the flight deck. The rest followed within seconds of each other. These messages were coupled with an audible alarm. The flight team in Mission Control also saw the warnings and calmly discussed the problem with the astronauts. The fault messages indicated a loss of pressure on the left main landing gear tires. Husband and McCool, along with Mission Control specialists on the ground, looked in the troubleshooting section of their shuttle manuals for these messages and reviewed the information to correct the problem.

But there was no solution to this problem. The superheated atmospheric gases had seeped into the wing and were melting the sensors, along with the landing gear and tires.

The crew had trained for almost every possible scenario on board the shuttle during their years at Johnson Space Center. Much of their disaster training involved an explosion on liftoff and how to escape the crippled machine on the launchpad or over the ocean. But the escapes they had trained for were at much lower altitudes, 36,000 feet and below. At the point of this emergency, they were at about 200,000 feet, or 38 miles up. The highest successful skydive at that time was from

135,890 feet—or 25.7 miles. But the *Columbia* crew had no way to eject from the spacecraft, even if they had known how dire the problem was.

The failure the crew saw was familiar, although slightly different from what happened in training. Back then, it was a circuit breaker trip that resulted in one-half of the tire pressure sensors going off. But the failure during the accident involved all the tire pressure sensors on the left main gear only.

In the minute before what NASA calls the "catastrophic event," one crew member on the mid-deck was not yet wearing his helmet, nor was he completely buckled into his seat, because he had been burning trash. Three crew members were not wearing gloves. For the six crew members who were wearing their helmets, they could not, per NASA protocol, keep their visors down because doing so results in high oxygen concentrations in the cabin as it is vented from the flight suits, which is a fire hazard. Astronauts have said the bulky gloves can inhibit the performance of even simple tasks. And astronauts have also reported that "the cabin stow/deorbit preparation timeframe is so busy that sometimes crew members do not have enough time to complete suit-related steps prior to atmospheric entry." In other words, they're too busy to put on and test all aspects of the flight suit before reentry. In the case of *Columbia*, though, the flight suits would have offered little to no protection for what would come next.

At this point, Mission Control reported nothing out of the ordinary in control or range of the ship.

At 7:58:48, the crew began a call to Mission Control, but the transmission was broken up in static and not repeated. However, that was common during descent as the shuttle's antenna changed angles toward different transceivers located throughout the country.

At 7:59:06, the "LEFT MAIN GEAR DOWN" lock sensor switched to "ON," even though, to their knowledge, the landing gear door was closed and the gear was locked in the up position. NASA officials are sure that the crew was attempting to diagnose the problem, particularly since it involved the same landing gear as the tire pressure messages and, thus, could have been indicating a dangerous landing gear situation.

At 7:59:29, all three reaction control system jets, which fire individually and intermittently throughout reentry to help stabilize the flight path, began firing simultaneously as the crippled wing began severely affecting the shuttle's flight.

At 7:59:32, Mission Control tried to communicate with Commander Husband regarding the landing gear issue. Mission Control heard the last words from *Columbia* when Husband tried to respond and was cut off midsentence: "Roger, uh, but . . ." In that instance, they thought nothing of the breakup of the communication. Again, gaps are typical during reentry. But within minutes, everyone in the large room in Houston knew this was more than a minor problem. There was a complete loss of signal from *Columbia*—no more voice communication from the astronauts and no more data communication from the thousands of sensors on the ship.

"I didn't expect this bad of a hit on comm," said astronaut Stephanie Wilson, referring to why she couldn't communicate with the astronauts. It was her job that day to talk with them during their descent.

At 8:00 A.M., *Columbia* should have been streaking across the southwestern half of the United States, leaving double sonic booms in its wake, fifteen minutes from the landing strip at Kennedy Space Center. Instead, the ship was breaking apart above Texas and Louisiana.

In the final forty seconds of flight, the crew knew something was going horribly wrong when they saw the lighting and horizon line change as the nose of the shuttle lifted up. The g-forces also increased from less than 1 to more than 3, far different from that of a normal flight. The pressure pushed the crew back into their seats and caused nausea, dizziness, and disorientation. The ship was 181,000 feet up, traveling at Mach 15—34 miles up and at least 11,418 miles per hour.

At 8:00:03 A.M., Houston time, either Husband or McCool returned the orbiter autopilot to the AUTO mode. At least one of the men was still mentally and physically capable of processing information and executing commands, an indication that the orbiter still had not lost pressurization. A panel was found among the shuttle debris during recovery, with switch positions showing that McCool had attempted to recover *Columbia*'s hydraulic systems and hydraulic pressure. While that action was not on the emergency checklist, NASA officials said it showed McCool's thorough knowledge of the shuttle systems as he worked to restore control. He spent the last moments of his life trying to save his crewmates and the priceless ship.

Below McCool on the mid-deck, one crew member still had not settled into his seat or finished strapping in. Only the shoulder and crotch straps appear to have been connected. The normal sequence for strap-in is to attach the lap belts to the crotch strap first, followed by the shoulder straps. The lap belts hang down between the closely spaced seats, and it would have been difficult for him to grasp the strap due to the motion of the orbiter, which may be why only the shoulder straps were connected. This same crew member was the only one not wearing a helmet. He was not named in the report.

At 8:00:18 A.M., space shuttle *Columbia* broke apart.

NASA was able to pinpoint the exact time because all power was lost to various instruments and experiments on board at that time. None of the astronauts had time to put their gloves on or pull down their visors.

The downstairs mid-deck broke apart first, causing an immediate loss of pressure and rendering those three astronauts immediately unconscious and unable to breathe. Then the upstairs flight deck and forward sections of the shuttle broke off, and those four astronauts lost consciousness as well. NASA's medical findings show that the crew could not have regained consciousness.

With the incapacitated astronauts still strapped in, both the mid-deck and flight-deck sections began to tumble. At some point, the astronauts' bodies became separated from the ship and tumbled to Earth. Remains from all seven astronauts were eventually found in the recovery effort.

According to NASA, the ship did not catch on fire or explode. It simply broke apart.

WFAA television station photographer John Pronk was in a Dallas park, hoping to shoot what is known in the television news business as a "beauty shot" of the shuttle's reentry as it flew overhead—just a minute or so of video of something pretty. He saw a bright light in the clear, blue morning sky, and he began rolling. Pronk's video captured *Columbia* as it disintegrated and turned into a half dozen bright lights and contrails streaking across the north Texas sky. His video was later referred to in the *Columbia Accident Investigation Board* report as confirmation to NASA of what had happened.

Lisa sat with her ten-year-old son by her side. He watched the television with her that morning, holding her hand, as the falling pieces of *Columbia* began to look like a meteor shower 230 miles to their north, near Dallas.

Nowak could hear Mission Control calmly but desperately trying to get Husband's or McCool's voice back on the intercom.

"*Columbia,* Houston, UHF comm check," one man in Mission Control said. There was no response. "*Columbia,* Houston, UHF comm check," he said again, to no avail. He would try again several more times.

As the seconds wore on, Flight Commander Leroy Cain seemed calm, but there was panic in his eyes, and his hand was over his mouth. He would later tell the television program *Nova* that he thought of the foam hitting the shuttle on takeoff and began to get a sinking feeling.

A fighter pilot, who was supposed to escort *Columbia* to Kennedy Space Center, reported that *Columbia* was late by one minute—an eternity in a program that times everything down to the second. And that, Cain said, was when he knew. About two minutes later, he moved to a contingency procedure. He ordered that the doors to Mission Control be locked and that everyone gather all the data from their computers. By now, many people in the room were on their feet, silently hoping or praying that their worst fears had not come true.

"You folks listen up on the flight route—no phone calls off-site outside of this room," Cain ordered. "Our discussions are on these loops and on recorded loops only."

In Norwood, Texas, nine-year-old Eli Connor was on his front porch when he heard repeated sonic booms, one after another. Pieces of the space shuttle *Columbia* began flying through the Texas sky and slamming into his grandfather's farm, along with the farms of their neighbors.

"I heard the first boom and I thought it was a plane going through," Eli told a reporter from the *Telegraph* newspaper out of London. "Then white turned to yellowish-orange and it was like a cannon going off. Every time it boomed again there were pieces coming down towards me."

More fiery pieces began raining down on the small community, some spinning through the air and sounding like a helicopter flying by. A helmet lodged into the ground between two oak trees.

As search teams made their way across the countryside, forming a line and walking across fields and through woods, they began finding pieces of the astronauts' bodies.

Mark Kelly wrote in his autobiography that he knew his astronaut classmates were already dead when he rushed to Johnson Space Center that morning. He was sent into remote parts of Texas to recover the bodies of his friends.

"An FBI agent and I had to take dirt roads and then go off into the woods. We recovered the remains of Laurel Clark, my very good friend and a U.S. Navy flight surgeon, on one of those dirt roads in Hemphill . . . Another search party found the remains of Columbia's pilot, Willie McCool, in a wooded area not far away. I immediately went to the scene to help with the identification and to make sure things were handled appropriately. On the second day, the FBI agent and I recovered the body of Dave Brown, a Navy captain who had been conducting science experiments on the mission. I spent two hours in the woods, sitting next to Dave's body, while the owner of the property cut back trees to get a larger vehicle to our location. I'd wished Dave well before the mission. Now, sixteen days later, all I could do for him was to make sure that his remains could be returned to his family."

At the Nowak house, Lisa sat in disbelief with her son, according to interviews she later gave.

"Mom, I know this is tough," she later recalled her son telling her, sounding more like a man than the ten-year-old boy he was. "But I still want you to go."

He insisted that she mustn't give up. She spent days and weeks and months afterward with another woman's son—astronaut Laurel Clark's boy, Iain.

Lisa was assigned to the Clark family as a "casualty assistance officer."

"She did everything," Laurel's husband, Jonathan Clark, a former NASA flight surgeon, told *Texas Monthly* magazine. "She went through everything: Navy paperwork, finances, bills, bank accounts. She took care of Iain during the months afterward. She saw what it was like to lose one of her best friends and for Iain to lose a mother. And the thing is, while Lisa is doing this, she is not at home with her kids. She has two very young children, but she is here twelve to fourteen hours a day under the most difficult circumstances. I have to think it was hugely stressful."

Her heart ached for this young man who had lost his mother and couldn't understand why. She treated him as she knew Clark would have treated her own son. And she grieved for one of the only women to whom she had ever really grown close.

8

★ ★

She Speaks

"I really didn't want or mean to hurt anybody," Nowak said softly to Detective Becton. "I just really needed to talk to her tonight."

Her reasoning didn't make sense to Becton—unless Nowak didn't want to talk with Shipman. Unless Nowak actually wanted to hurt Shipman, instead . . . maybe even kill her.

"I wanted to just get in her car and talk to her, and [she was] understandably scared, and I, I just . . . there was this tiny chance, I needed to talk to her. And she's going away, I was like . . . it was stupid."

"Where do you live?" he asked her.

"Don't you have that?"

"No, actually I don't. Where do you live?"

"Why do you need to know?"

"Because I'm going to make a point with that."

"What is it?"

"You need to answer my question first," he said.

"Tell me the point first."

Oh, for goodness sake, he thought. *Is she kidding? Why was she screwing with him on such a simple question? Again, she was ten steps ahead of him and maneuvering to the eleventh step.*

"You need to answer my question first," he said.

"I live in Houston."

"OK, you're here in Orlando, which means you have to fly here, am I correct?"

"There's more than one way to get here."

Why wouldn't she just tell him?

"You could have driven here, or you could've taken the Greyhound. I doubt that you took a bus," Becton said. "But you had to come to Orlando from Houston. And I see you have three children. So, that means your children are still in Houston. So, you left your children to come to Orlando just to chitchat?"

This was actually making him a little mad. Yes, she was a rocket scientist and he was just a cop, but have some respect.

"I'd greatly appreciate it if you would not insult my intelligence by the cockamamie story that you just gave me. I would rather you spit in my face as if you were no better than somebody out in the ghetto than insult my intelligence. Either you want to be up-front and you want to tell me your side of the story, or you don't—it's that simple . . . You're toying around with me."

"No, actually I'm not," she said.

"You are," he fired back. "'Cause what you just told me, 'I'll just sit here and talk to her and sit in my car and talk to her.' Bull!"

I'm going to use the mommy card, he thought, *tug at her heart a little bit to see if that would open her up.*

"You don't leave three children in Houston. I mean, where are your children? Is there somebody watching your children? Hopefully, somebody you trust because you're in Orlando and they're in Houston, and if your children needed you like my daughter needed me this morning, you're not gonna be there for them."

He watched her closely.

"I think we're at a point right now where all we can do is minimize the impact of what's occurred," Becton said.

"How do we do that?" she asked.

"I don't know. It depends on how honest you're going to be with me and what you tell me."

"Everything I told you is the truth," she said.

"So you want me to just state that an astronaut with NASA left her three children in Houston to come to Orlando, to speak to another person."

Her son would be fifteen tomorrow. He had been her rock through so much these last few years, her best friend, really. And her adorable twin girls. They were five and getting into everything and making her smile so much and asking so many questions.

"They're with their dad right now, like they are every other weekend."

Thoughts seemed to tumble about Richard, their time at the Naval Academy, their wedding, her son's birth, the miscarriages, the girls' birth, her flight in July, the separation two months ago. They were trying to work things out and be cordial for the children's sake.

"Is your divorce final, or are you just separated?"

"Separated," she mumbled.

"Is this something you wanted or he wanted?"

She mumbled again . . . Becton thought she seemed to become overwhelmed by the sadness of her failed marriage.

"Did he break your heart?" Becton asked. "You didn't like him?"

"My husband is the only person who broke my heart," she said. "Gradually. Over the years, I guess. It happened over time. People grow apart sometimes."

Maybe she wasn't the only one cheating, he thought. *Maybe Richard Nowak found someone else first and that's why she turned to Oefelein. I know that feeling of betrayal.*

"It happens," Becton said, thinking about his first wife, the one he walked in on, cheating on him with their attorney, who had also acted as the couple's financial adviser.

"You don't believe me," she said, laughing a little.

"Oh, I believe you," Becton nodded. "It could happen. Do you think he got jealous over your career?"

"No."

"Was he proud of you?"

"I think he was."

Becton talked about watching her launch on the Fourth of July, from a spot along the Banana River. She said she thought she knew where they had been that day.

"You have a lot going on inside of you," he said. He knew people attacked others for one of three reasons—drugs, money, or a broken heart. "It's either a lot of pain, a lot of anger, or it's both. And right now, you're bottling it up a lot. Hear me out, 'cause, you know, you and your husband grew apart and knowing how the common person sees you and sees this, you would have thought that your husband would have been extremely proud of you."

She said something that he couldn't quite make out. He decided to go back to the task at hand.

"I cannot prevent you from going to jail, and I can't prevent that you, your name, and her name from being entered into any paperwork," he said.

"What if she drops the charges?"

Wow—she really was way ahead of him on this. Hardened criminals never asked about charges being dropped—they knew better.

"I don't think she's gonna drop the charges."

"But if she did?"

Many times, especially in inner-city neighborhoods with a code of silence, victims don't want to cooperate. That's why police and the state attorney's office can pursue charges without them.

"I want to press charges and I'll be a witness," Becton told her. "I doubt very seriously she's gonna drop the charges."

"Maybe if you could let me talk to her," she pleaded. It would be a request she would make over and over throughout the rest of the morning. "Let me talk to her."

"No."

"Why not?"

"No, absolutely not, 'cause I don't know what you're capable of doing right now."

9

★   ★

Billy-O

Billy Oefelein was an Air Force brat, the younger son of Randall and Bil-lye Oefelein and little brother to Randy. He was born in March 1965 in Ft. Belvoir, Virginia, a cool spring day sprinkled with some rain in the Washington, D.C., area.

His dad's job took the family to Alaska when Billy-O (as everyone calls him) was ten years old, and that is the place where he now feels most at home—hiking trails through lush meadows, canoeing crystal-clear rivers, climbing mountains in the summer, and in the winter run-ning snowmobiles across iced-over lakes and flying out into the wilder-ness that still looks much the same as it did when Jack London wrote *Call of the Wild.*

As he told NASA in an interview before his shuttle flight in Decem-ber 2006, he joined the Boy Scouts in elementary school and began camping and hiking then. He was in his early teens when he realized that the best way to see Alaska was by plane, and so he began taking flying lessons. He did his first solo flight in a glide plane at fourteen and got his private pilot license at nineteen.

"I think the people that stay [in Alaska] see what other people see as obstacles—no roads and all that—as opportunities to get out and explore, develop new roads, or develop new communication systems or somehow connect the remote parts of the state, to develop new technologies or employ existing ones like airplanes to connect the state," Oefelein said. "I think that just makes you more of an explorer, when you see things like that, and you don't see them as obstacles but opportunities."

Just months after the *Columbia* disaster in 2003, Oefelein met with a group of school students at the Kenai Boys and Girls Club in Alaska, showing them pictures of glaciers in south-central Alaska, taken from the International Space Station.

"When I was your age, it was pictures like this that kind of got the fire going in me," the *Peninsula Clarion* reported him saying. In an interview with NASA he also said, "I didn't grow up wanting to be an astronaut. I just wanted to fly airplanes and explore. I always liked math and science."

Oefelein's ultimate determination to not only live but to live the life of an adventurer came about a month before his eighteenth birthday, four months before his high school graduation.

In an essay for a book for the Forget Me Not Foundation, a group dedicated to preventing drunk driving in Alaska, Oefelein said he remembered being so very, very cold one February night in 1983. He was strewn on the pavement near his home in Anchorage, not able to remember that he had just been hit by a drunk driver as he pushed his stalled car off the roadway.

"The drunk got out of his truck and glanced nonchalantly at my motionless body, now lying in the middle of the road," Oefelein wrote, recalling what witnesses had told police: "[The drunk] surveyed his truck, which was stuck in the snow, and put it in four-wheel drive. He then got back behind the wheel, freed himself from the snow bank, and sped away."

A taxi driver who had witnessed everything radioed his dispatcher to call the police and then followed the drunk driver, who was arrested. Several Good Samaritans stopped to tend to the badly injured Oefelein.

Rescuers rushed the handsome Boy Scout to the Air Force base hospital at Elmendorf. Oefelein drifted in and out of consciousness, not realizing he was in shock. He had lost blood and was desperate to get warm. Relief came when nurses began piling heated blankets on him in the emergency room.

As his parents waited outside his room, two doctors and three large orderlies worked to correct his broken leg—his right leg from the shin down was facing the wrong direction, held onto his body by just a few muscles and tendons. Both his tibia and fibula were broken and poking out of his skin. In what he describes as the most physical pain he has ever felt—and without the comfort of anesthesia—four men held him down as the doctor twisted his leg back into place with one swift, excruciating move. The good news was that the doctor thought he would "probably" walk again.

It took three surgeries and sixteen months of physical therapy to fix the damage, although it rendered his right leg half an inch shorter than his left and gave his walk a little bounce.

In court, Randall Oefelein asked the judge to break the leg of the drunk driver as punishment for hitting his son and leaving him to die. But the judge said he couldn't do that and, instead, sentenced the man to the maximum of eleven years in prison. It was just his latest in a collection of DUI convictions.

"Unlike me, the man who ran me over and left me for dead walked out of the courtroom unassisted," Oefelein wrote. "He served four of his 11-year sentence in prison, spending the last seven on parole."

Decades later, his mom, Billye Oefelein, is still angry at the man who hit her son.

"How could someone just leave my son alone in the middle of the road to die?" she asked. "How could he get out of his car, see that he'd hit a kid, see that kid bleeding in the middle of the road, then get back into his car and drive off? How could he do that to a kid?"

Oefelein graduated from West Anchorage High School four months after the accident and had plans to go to Oregon State University (OSU) for college. But he spent his freshman year of college at the University of Alaska while he continued to rehabilitate. He eventually headed to OSU and graduated in 1988, earning a degree in electrical engineering.

That year, he also married Michaella Davis, whom he began dating when she was in high school, according to her mother in a 2007 *Houston Chronicle* article.

Oefelein left the temperate climate of the Northwest and headed to Aviation Officer Candidate School in tropical Pensacola, Florida. Oefelein received his commission as an ensign in the United States Navy in 1988 in the base chapel. Like most candidates, he gave their drill instructor a token of thanks, handing him a silver dollar as he stepped up for a salute and handshake.

Oefelein entered flight training in Texas in 1989 and earned his wings in September 1990—just after Saddam Hussein invaded Kuwait. He then reported to Marine Fighter/Attack Training Squadron 101 at Marine Corps Air Station El Toro, California, as a "Sharpshooter" for initial F/A-18 Hornet training.

The F/A-18 Hornet is the workhorse of the Navy and Air Force, with the now familiar V-shaped tail and leading edge wing extensions. They can take off and land on military runways or on an aircraft carrier and fly up to Mach 1.8—1,190 miles per hour.

Oefelein wore a green or desert camo flight suit, with his name on a patch over his left breast. Fingerless gloves came in handy inside the cramped cockpit—a man weighing over 200 pounds might not be comfortable flying this jet fighter. At about five foot eight and a trim 160 pounds, Oefelein was a perfect fit for the jet.

He maneuvered the aircraft using a joystick between his legs and looking at an instrument panel about 2½ feet tall and wide, a bent arm's length in front of him. He wore a helmet with a black visor that covered his eyes, and a mask with a rubber tube—snaking down to the oxygen tank—covered his mouth and nose.

Inside the cockpit, it sounded like the dentist's suction machine times one thousand as the jet traveled at supersonic speeds, climbing straight up or shooting straight down. It could fly upside down and do loops—any maneuver Oefelein wished. It could also fly at low altitudes, skimming hills and treetops to avoid radar detection.

Upon completion of training in 1993, Oefelein was assigned to Strike Fighter Squadron 146 at Naval Air Station Lemoore, California, from where he made overseas deployments aboard the aircraft carrier USS *Nimitz* to the Pacific and Indian Oceans, and the Persian Gulf in support of Operation Southern Watch, which enforced the no-fly zone in southern Iraq following the Gulf War.

Operation Southern Watch began on August 27, 1992, after Saddam Hussein's forces repeatedly bombed and strafed Shi'ite Muslims in southern Iraq, with heavy fighting in Basra. The United Nations Security Council demanded that Iraq "immediately end this repression and express[ed] the hope in the same context that an open dialogue will take place to ensure that the human and political rights of all Iraqi citizens are respected." Forces from Saudi Arabia, the United States, the United Kingdom, and France participated in the operation.

Military engagements in Southern Watch were frequent, with coalition jets often shot at by Iraqi forces. Between August 1992 to early 2001, coalition pilots flew 153,000 sorties over southern Iraq. While on board the USS *Nimitz*, Oefelein flew twenty-nine combat missions over the skies of southern Iraq.

Like other pilots on board, Oefelein would salute the *Nimitz* deck crew, circle his joystick to maneuver the wing flaps, and then catapult off the short part of the aircraft carrier's deck, taking about two seconds to reach flight speed. He could be jolted in his seat when the jet dropped off the lip of the carrier, but then took off in flight.

Southern Iraq is a desolate place, with just a few towns peppering the desert. Basra is its best-known city; located between Kuwait and Iran, it is the second-largest city in Iraq. It is also the country's main port—from where millions of barrels of Iraqi oil is shipped daily—and also the port from which Sinbad the Sailor set out on his famous journeys. Temperatures in southern Iraq can reach 122 degrees Fahrenheit. On the lonely highway between Baghdad and Basra, on some days all that can be seen are a few wild camels crossing the road.

Oefelein wrote about his time during Operation Southern Watch on his Adventure Write website. "The idea was to fly like someone was trying to kill you, especially around the boat."

According to the U.S. military, during 1993, U.S. forces destroyed several Iraqi radar sites in the no-fly-zone area south of Baghdad, along with Iraqi Intelligence Service headquarters in Baghdad, in response to an assassination attempt on former president George H. W. Bush in Kuwait. (On April 13, 1993, fourteen men believed to be working for Saddam Hussein smuggled bombs into Kuwait, planning to assassinate former president Bush by exploding a car bomb during his visit to Kuwait University three months after he had left office. The plot was foiled when Kuwaiti officials found the bomb and arrested the suspected assassins.)

Landing a supersonic jet on an area about the size of a football field took precision and practice. Oefelein had more than two hundred carrier-arrested landings on the USS *Nimitz*. He would slowly come in, almost appearing to drift just above the edge of the flight deck as he dropped the jet's tailhook, which caught a cable and halted the aircraft so it wouldn't slide off the other end of the giant carrier.

In 1994, while assigned to VFA-146, he attended the U.S. Navy Strike Fighter Tactics Instructor program in the U.S. Naval Air Station Miramar, north of San Diego (now a U.S. Marine airstrip). It is known by its more familiar name—the famed TOPGUN School—attended by Tom Cruise's character in the movie. Oefelein was assigned as the squadron air-to-air weapons and tactics officer.

According to the Navy, the TOPGUN program develops, refines, and teaches aerial dogfight tactics and techniques to selected Naval aviators and Naval flight officers, who return to their operating units as surrogate instructors. It was designed to train already experienced Navy and Marine Corps aircrews at the graduate level in all aspects of strike-fighter aircraft employment, including the current world threat for air-to-air and air-to-ground missions.

The TOPGUN course has changed over time. In the 1970s, it was four weeks long; in the 1980s, it grew to five weeks. At the time of Oefelein's training, it was more than two months long. It was developed during the Vietnam War era, following losses during the famed Operation Rolling Thunder (the bombing of North Vietnam), which lasted from March 1965 to November 1968. Rolling Thunder ultimately saw almost one thousand U.S. aircraft losses in about one million flights. One of the planes shot down was that of John McCain, the son and grandson of two admirals. He was held as a prisoner of war in North Vietnam for six years and endured torture that left him permanently disabled.

At TOPGUN, Oefelein and his crewmates took eighty hours of lectures and twenty-five flights in their F/A-18s that pitted them against TOPGUN instructors.

"When I went through TOPGUN, that was some of the most intense, yet fun, flying I have ever had the privilege of doing," Oefelein later recalled.

When finished with TOPGUN, Oefelein was selected for the United States Naval Test Pilot School at Naval Air Station Patuxent River, Maryland, and began the yearlong course in January 1995.

It was here that he first met and became friends with Navy Captain Lisa Nowak, who had a husband and young son. And he had a wife and two children, ages six and three.

The test pilot school provides instruction to experienced pilots, flight officers, and engineers in the theory, processes, and techniques of aircraft and systems test and evaluation. Two forty-eight-week courses are offered every year, with thirty-six students in each class, including pilots, flight officers, and engineers from all military branches. The school involves more than five hundred hours of classroom instruction and one hundred hours of flight time in more than fifteen different types, models, and series of aircraft.

"A lot of folks who do that test pilot work also went on to fly space shuttles," Oefelein said in an interview with NASA. "I started talking to a bunch of those folks and at that point it just seemed natural for me to go to the next phase and try to fly space shuttles. And, I just kind of found myself in a position . . . 'Hey, I can use all this background and test aviation and exploration and maybe help fly a space shuttle for NASA someday.'"

He was assigned to Strike Aircraft Test Squadron as an F/A-18 project officer and test pilot.

"Probably the hardest training I did in my Navy career was studying to become a Test Pilot," Oefelein wrote in a NASA blog. "There was a lot of math and science required, in addition to the precise flying and report writing we had to do."

In February 1997, he went back to the United States Naval Test Pilot School, this time as an instructor as his career continued on a successful upward trajectory.

In February 1998, he transferred to Carrier Air Wing 8 as part of a squadron, at Naval Air Station Oceana, Virginia, where he was assigned duties as the strike operations officer. He also completed a master's degree in aviation systems at the University of Tennessee's Space Institute.

And then he got word that year that he had finally been selected for the astronaut program. He was thirty-three years old.

"A lot of people think the first thing you're going to do is fly on the space shuttle," he is quoted as saying in the *Peninsula Clarion* in 2003. "That's not exactly how it works."

He explained to the students at the Kenai Boys and Girls Club that there is extensive training involved in becoming an astronaut—sometimes fun training.

"Who here likes to play video games?" he asked, in response to which every hand in the crowd went up. "One of the things you get to do as an astronaut is you get paid to play video games."

Of course, he was referring to the shuttle simulators.

Oefelein reported to Johnson Space Center in August 1998. He was initially assigned technical duties in the Astronaut Office Advanced Vehicles Branch and Capsule Communicator, known as CAPCOM, talking to astronauts on board shuttles and the International Space Station from Houston.

In 2000, he went on two important training exercises. A visit to Star City, Russia, taught him "how the other guys do spaceflight." And at New Year's, he was in Antarctica, searching for meteorites around the Darwin Mountains. He and his team dug up about eight hundred "extra-terrestrial rocks."

Late on a Friday afternoon in the spring of 2002, Oefelein was working at CAPCOM for a space shuttle mission when he got a phone call to go to the corner office—his boss's office.

"Well, this is going to go one of two ways: It's either going to be really good or really bad," Oefelein recalled in a NASA interview. "I didn't think I'd screwed anything up, but maybe I did . . . but I didn't think it would go too bad."

But as he walked down the hall, he had another realization.

"So I kind of had in back of my mind that maybe this is the call that you're waiting for," he said. "Sure enough, I go in there and there's the chief of the office at the time . . . They asked me, 'Hey, you still interested in flying in space?' I said, 'Sure.' And then they said something about [the mission] and talked for another 30 minutes. I caught about five or six words of that."

Oefelein was thrilled and was allowed to go home and tell his wife, Michaella, and their children, but no one else.

"They were pretty happy. It was good news," he said. "It was a little while coming at that point, and everybody knows that I work hard and my family made a lot of sacrifices. So, it was good."

Within nine months, though, everyone's dreams were put on hold following the *Columbia* tragedy, when all shuttle flights were grounded as NASA explored the cause.

His training, though, continued. In 2004, Oefelein went on a survival training exercise with five other astronauts, including Lisa Nowak. Later that year, he climbed Mount Rainier for a friend's wedding, and did team-building and leadership exercises at Knight Island, Prince William Sound, Alaska, including kayaking as a wounded ankle healed.

The next year, he was off to Mexico for more mountain climbing—this time Mount Orizaba, the tallest peak in Mexico.

"Some days you beat the mountain . . . some days it beats you," he wrote on his Adventure Write website.

Robber's Roost Canyon in Utah was the site of a ten-day, STS-116 crew team-building exercise, with the astronauts exploring canyons. The lesson here: "Learn to trust your crew before you leave the planet."

10

★ ★

New Moon

One year after *Columbia* sailed above Earth on its final, doomed flight, Lisa Nowak and Billy Oefelein went on wilderness training in a remote area of Canada, north of Quebec, under the January new moon—only the stars were visible. For Nowak, after a year of taking care of Iain Clark and handling all the legal paperwork for the Clarks as their NASA family casualty assistance officer, the exercise in the Canadian wilderness was a respite from the grief and strain in Houston.

From January 19 to 29, 2004, Nowak, Oefelein, and four others—Dominic Antonelli, Christer Fuglesang, Dmitry Kondratiyev, and Julie Payette—underwent four days of training by the Canadian military and then were dropped off in the snowy, hostile backcountry of Canadian Forces Base Valcartier, according to newscientist.com. "There, they moved camp each night, trekking a total of more than 20 kilometres"—or 12½ miles—no small feat in snow and freezing temperatures.

Several years later, Oefelein wrote on his webpage Adventure Write: "What do you get when you mix a Swede, a Russian, a Canadian, and three Americans together on a daily mission to move camp in temps as low as 40 below? Quite a dog team!"

According to NASA, this training is about much more than building a campfire and shelter or living off of insects and small animals. It can show the true character of an astronaut and whether or not they will make a good team member in the stressful, inescapable confines of space.

Why train in a frozen tundra when space shuttles launched from and landed in sunny Florida? During the space shuttle era, astronauts who enjoyed an extended stay on the International Space Station (ISS) sometimes had to return on a Soyuz spacecraft. All astronauts currently come and go on the Russian rockets, which look like something straight out of the Apollo era. The spacecraft launches from Baikonur

Cosmodrome, located in a remote area of Kazakhstan, and lands nearby between Dzhezkazgan and Lake Balkhash, on deserted steppes that can be awash with snow or as barren as a Martian landscape. The annual temperature averages 55 degrees Fahrenheit but can reach minus 40 degrees in winter and 113 degrees in summer.

Landing a Soyuz is an imprecise science because of the use of parachutes. The wind can pick up and carry the trio inside the capsule nearly twenty miles off course. If something goes wrong during reentry, the bell-shaped capsule can plummet more than 370 miles away from the landing target.

A crash in this landscape could "leave the humans on board to fend for themselves for a while," NASA wrote in a press release in 2004. "Knowing how to take care of their basic needs would be invaluable."

The Canadian winter-survival-skills missions "include being able to construct a shelter for protection from the elements, build a fire, obtain food and water to supplement the existing supplies, and deal with first-aid emergencies. That's the physical side to survival training," NASA stated.

Michael Lopez-Alegria, an astronaut and ISS crew operations manager at Johnson Space Center, said instructors hope to instill self-care and self-management skills, to develop teamwork skills, and to strengthen leadership abilities.

"It's important for these people to know the extent of their capabilities under extreme conditions," Lopez-Alegria said. "They're going to have to be able to get along in an isolated environment for long stretches of time. The Space Station and the Space Shuttle might be a bit more comfortable than the wilderness, but in many respects the end result is the same: you're stuck with these other people, and your survival depends on everyone working together."

Lopez-Alegria said that during his survival training, he found that the simple task of brushing his teeth became a big chore because it involved finding water and setting up any makeshift toiletries, then cleaning up after himself.

"I was struck by the planning and forethought required for such a simple self-management task," he says. "That made a big impression—bigger, perhaps, than the larger jobs of staying afloat at sea or keeping warm in the snow."

"Right away they're going to be learning about the priorities of survival," Canadian Master Corporal Jonathan McArthur said in a 2017 globalnews.com article. "These are self-aid, first-aid, fire, water, shelter, signals and food."

It is unclear how much time alone Nowak and Oefelein spent on their survival training, but by Oefelein's account, their romance began in 2004. Certainly, there were nights by campfires.

Their first task after being dropped off was to build a shelter from tree limbs fashioned into a lean-to. Survivalists in the training program are often left with limited supplies that included a coffee tin, wire, a sleeping bag, rations, a saw, and half of a parachute.

"The biggest problem is getting cold from the [frozen] ground," said Corporal Jonallan Lindley-Scott in the globalnews.com article. "When you have a nice, thick, pine bough bed, it's just as good as any mattress at home."

Wilderness survival expert Mors Kochanski is brought in to teach pilots the importance of knowing how to start a fire and dressing for the weather. Kochanski explained to globalnews.com that most people who die in the wilderness do so within the first thirty-six hours of being stranded.

"The two things you have to do to survive is get enough sleep, which is two hours of rapid-eye-movement sleep, and drink enough water," Kochanski said. "Without you paying attention to those things, you succumb very quickly."

The next task is to build a massive signal fire that rises above the trees so that rescue helicopters can see it. They're taught to pile evergreen limbs like a teepee, leaving a hole in the bottom for oxygen to come through.

Nowak and Oefelein are seen in several photographs, standing side by side, smiling, wearing olive-drab green arctic wear. In one shot, they seem to be getting ready to head out, with all their gear piled in front of them, including snowshoes. In another shot, standing in the back behind everyone else, their arms are angled toward each other.

Despite the shuttle fleet being grounded, the astronauts continued to train in Houston. In 2004, after waiting since being selected as an astronaut in 1996, Lisa was assigned as a mission specialist to the crew of STS-121.

As *Texas Monthly* reported: "Suddenly her workweek was more like 70 hours. In addition to her job, she had to take care of her twin toddlers and her teenaged son. It was more work than could reasonably be done by a person getting a good night's sleep."

"I tell young girls that the best jobs in the world are being an astronaut and being a mother," astronaut Eileen Collins told the monthly magazine. "But it is not always easy. The difference is, for a woman, the husband usually works, so you end up doing all the work at home. I will tell you that sometimes I went on two or three hours of sleep. I don't like to push that or brag about that, because it could be unsafe. If I am going to fly, I'm going to make sure I get more sleep. But sometimes I just had to stay up because the job didn't get done. When do you answer your e-mail? You do it at home, because you have no time at work."

Jonathan Clark, Laurel Clark's widower, agreed.

"It's almost incomprehensible how much stress there is," he told *Texas Monthly*. "First, there's the fact that you're doing a high-profile job that requires a lot of travel. There are 80-hour weeks. And it's different for a woman. She takes care of the kids differently than a man. Most men are just not around, but she does not have that option. My perspective is that it puts immense stress on a marriage. For us it certainly did. The only way you cope with it is through incredible tolerance and flexibility."

In late 2004, Nowak and her husband decided that they should separate and divorce.

"They agreed to do this after she returned from her shuttle flight," a 2007 psychiatrist's report states. But "the flight then had multiple delays of over a year."

In the spring of 2005, workers readied space shuttle *Discovery* and Nowak and her crewmates for the Crew Equipment Interface Test, or CEIT, something each crew does about two months prior to their flight. NASA documents show it was developed to prepare astronauts for their missions in space while still here on Earth, and it is designed so the crew and staff can have full access to every area of the shuttle. It gives astronauts hands-on training with the actual tools, equipment, and hardware they'll use in orbit. For Nowak and other first-time astronauts, it was their first encounter with the shuttle in which they would fly.

After the training was finished, Nowak told crewmates Mike Fossum and Mark Kelly that she wouldn't be joining them on the trip back to Houston. Instead, she said she was staying in Cocoa Beach for a few days to look for a reception hall for her flight party. But that wasn't entirely true.

When Fossum and Kelly arrived at the airfield to take off for Houston in their T-38, they noticed a flight plan written into the log just before they arrived: pilot Billy Oefelein and passenger Lisa Nowak were headed to Naval Air Station Key West. Fossum would later say it was peculiar because Nowak had no reason to deceive them.

Oefelein also showed up a day early at NASA's Ames Research facility, one of ten NASA field centers, when Nowak, Fossum, Kelly, and Commander Steve Lindsey were there for a tour and to meet researchers associated with their mission.

In Houston, Fossum said, he had once spotted Nowak and Oefelein leaving a party together. And several colleagues said they had seen Nowak chatting with Oefelein in his office several times or seen the pair working out in the astronaut gym together or at the same time.

In the fall of 2005, *Texas Monthly*'s S. C. Gwynne was allowed to observe some of the grueling training the STS-121 crew conducted. They spent twelve to sixteen hours a day honing their skills and reflexes in the most extreme, life-or-death scenarios. Sitting beside Gwynne in the shuttle simulator was Mission Specialist Lisa Nowak.

"She seemed intense, focused, not at all relaxed, and not entirely thrilled that a journalist was sitting next to her," Gwynne wrote. "People have described her as shy, and she did seem that way, though she shared in the casual jocularity of the crew. Mostly, she struck me—with her hair pulled back into a tight ponytail, makeup-less and relatively cheerless—as all business. Maybe that was because the essence of simulator training is crisis."

Gwynne described instructors in another room creating rapid power failures during orbit reentry, leaving the crew scrambling to find a solution. It was, essentially, the same scenario her friends on *Columbia* had faced and died in three years earlier.

"Sometimes these problems are unintentionally fatal, and the crew is [theoretically] 'killed,'" Gwynne wrote of the training.

Some emails from 2006 show friendly, mainly work-related exchanges between Nowak and Oefelein about complicated space shuttle topics. Often, when Nowak learned something new, she would forward it on to Oefelein. She would also forward her schedule to him.

In February 2006, Oefelein secured a jet for training for Saturday, February 18.

"Hope you're having fun," he wrote. "Call when you can."

Before her July 4 flight into space, Nowak and Oefelein were seen together in the secure and quarantined Kennedy Space Center crew quarters—a place only a spouse and three guests can be and only briefly before a flight. One guest is allowed after a flight.

11

★ ★

The Launch

July 4, 2006

Lisa Marie Nowak and her crewmates woke up in the Kennedy Space Center (KSC) crew quarters at 4:38 A.M. on July 4, 2006—ten hours before liftoff of Space Transport System 121 (shortened to STS-121). Two previous launch attempts on July 1 and 2 had to be scrubbed due to typical Florida summer weather—thunderstorms with lightning—so the crew was anxious.

Theirs was the second "Return to Flight" mission that NASA was sending into space following the *Columbia* accident. Every facet of their mission would be watched by the American people, particularly the thousands who worked at NASA facilities all over the country. The astronauts' mission while in orbit was to photograph the underbelly of the shuttle, along with the wings' leading edges. They would test out the shuttle's robotic arm to see if astronauts could stand on its end to do repair work on the shuttle, if needed. Mission specialist and flight engineer Lisa Nowak would be at the forefront of those tasks as the lead controller of the shuttle's robotic arm and orbiter boom, along with the robotic arm of the International Space Station (ISS).

They would also test repairing shuttle tiles in the zero-gravity environment of space. Their mission had to be successful—as close to flawless as possible—to boost public and congressional support for a program in peril.

S. C. Gwynne wrote for *Texas Monthly* magazine in May 2007 that each space shuttle is an extremely fragile and "balky patchwork of far-flung technologies, many of which are a generation old. To get its three million pieces into orbit without exploding or disintegrating requires an astonishing amount of money and effort: as much as $1 billion and 20,000 workers *every time it blasts off.*"

NASA's website shows that, at Kennedy Space Center, the crew quarters are nestled inside the Operations and Checkout Building, a nondescript, square, five-story structure located off an access road a few miles between the visitors' complex and the launchpads. In its Apollo heyday, lunar capsules were checked out in this building before they were attached to the giant Saturn rockets.

The crew quarters are on one level, laid out like a hotel floor, with dorm-style living areas. On another level is the suit-up room, where astronauts put on the four layers of their space suits.

Decorated with Scandinavian-style furniture from the 1980s, the era when the shuttles first started flying, each of the twenty-three bedrooms of the crew quarters has a bed, wardrobe, television, clock radio, and private bathroom.

A staff of five people took care of the astronauts' needs, prepared meals, cleaned up quarters afterward, and disinfected every surface to ensure the crew's health. All staff had to pass physicals before they came into contact with the astronauts in this highly quarantined area. Astronauts were placed in quarantine for three days at Johnson Space Center and for four days at KSC prior to launch. One sick astronaut could ground and delay an entire mission.

"I treat them like my children," Delores Abraham, who worked in the crew quarters for five years, once told NASA officials. "I always told them they were my boys and girls."

In fact, Eunice Allen became a sort of den mother to the Apollo and shuttle crews, baking them homemade chocolate chip cookies and soothing any fears they might have. And she was part of their families, babysitting children while astronauts and wives went on extended vacations. Bruce Melnick and Mike McCulley both welcomed her into their homes.

Wendy Gorie, wife of astronaut Dominic Gorie, explained to NASA officials that the final day before launches, the crew gets together with spouses for a picnic at the beach and emotional good-byes.

There are several conference rooms in the crew quarters, including one that also serves as a dining area. One conference room has a large-screen television and a collection of DVDs—including, of all things, the movie *Apollo 13*. Failure wasn't an option for Lisa Nowak and this crew, either.

Looking around, it had the ambiance of a second home. There were photographs of astronauts' children, smiling and holding homemade cardboard shuttle replicas. Gurgling, cooing babies clung tight to mission emblems stuffed into their fat, little fists. There were even a few dogs and cows pictured.

"That way, they can see their families—you know, make them feel at home," Laura Lunde, a flight crew support specialist who managed the site, told NASA.

Family members were not permitted to stay in the crew quarters, and visits were limited to a spouse and three guests for only a few hours, provided they were cleared by a NASA doctor first.

And that's why it came as a surprise to crewmate Mike Fossum to see Nowak in the crew quarters hallway with Commander Billy Oefelein before their July 4 mission. He wasn't on their flight. He was not her husband, not related to her in any way. Although there had been whispers. And there was the not-so-secret flight to Key West the year before.

"This was odd because very few individuals other than crew members had reason to be in the area," Fossum later told investigators.

The crew ate a hearty American breakfast together, then gathered around the table to take a picture with a three-dimensional mock-up of the flight emblem—a preflight tradition. All were dressed in matching short-sleeved maroon polo shirts with the STS-121 logo on the chest.

By about 10:30 A.M., the crew moved to the suit-up room, four hours before T-minus zero. Before every Apollo flight, astronauts gathered here to put on their flight suits. Armstrong, Aldrin, Collins, Young, Crippen, Mitchell. All of them.

On this day, those historic names would be joined by Nowak, Commander Steve Lindsey, pilot Mark Kelly, spacewalkers Mike Fossum and Piers Sellers, robotics engineer Stephanie Wilson, and German astronaut Thomas Reiter of the European Space Agency.

The white flight suits of Apollo gave way first to silver suits for shuttle astronauts and then bright-orange suits—affectionately dubbed "pumpkin suits." The seventy-pound piece of equipment, technically called an Advanced Crew Escape Suit (ACES), is tailor-made for each astronaut from hundreds of precise measurements. It has its own communications, medical monitoring, and cooling systems, along with a layer to protect against g-forces, and is made by the David Clark Company of Worchester, Massachusetts.

Each astronaut began to get ready inside private rooms. The first item they put on was a "Maximum Absorbency Garment"—NASA-speak for an adult-sized diaper.

"The suit takes a long time to put on and take off and, well, you get the picture," astronaut Eileen Collins once said in an interview with NASA. In addition, it takes a while to get out to the launchpad, sit in the crew compartment, and wait for liftoff.

All of them were used to wearing it because all of them had trained in NASA's Neutral Buoyancy Laboratory—a swimming pool 40 feet deep, 202 feet long, and 102 feet wide. It holds 6.2 million gallons of water and offers the closest experience on Earth to walking in space. Astronauts are often in their bulky training suits and underwater for almost a full workday, with no time for a bathroom break.

During the Terminal Countdown Test—a practice for launch—for a 2001 flight, STS-121 Pilot Mark Kelly explained in his autobiography *Gabby: A Story of Courage and Hope,* that he not only used Huggies diapers, but he also usually utilized four feminine hygiene maxi-pads. On that practice launch, though, he stuffed only three pads into his Huggies.

"The test lasts for six hours and we can't change out of those pull-ups," Kelly wrote. "When we're on our backs for those hours with our feet up—the position astronauts are in for launch—the fluids shift in our bodies and our kidneys start working overtime. It wasn't long before I was filling the Huggies. And I quickly realized that the awkward positioning of the maxi-pads, combined with my mistake of being one maxi-pad short, was not helping the situation. I could feel the urine soaking into long johns on my legs, and then heading up my back. There was so much that over the next two hours, it defied gravity, working its way up my right leg, which of course was higher than the rest of my body. It traveled uphill, and even soaked the top of my right sock. By the time I de-suited a couple of hours later, I had ten pounds of wet diapers and maxi-pads to dispose of."

Kelly had flown once before, going up three months after the September 11 tragedy. He grew up in New Jersey but called Tucson, Arizona, home. His then-girlfriend, Gabrielle Giffords, was the U.S. congresswoman from that district. At the time of the launch, the pair had come under scrutiny because she chaired a panel that oversees NASA, and critics say Kelly had undue influence on Giffords. She said she was a fan of NASA well before he came along.

At forty-two, Kelly sported a mustache and kept his thinning hair at a close crop. Not to be outdone, his twin brother, Scott, was also a space shuttle pilot and commander for NASA. As Kelly described in his autobiography, the pair grew up locked in a real-life rock-'em, sock-'em battle, tumbling throughout their police-officer parents' home, breaking furniture, and knocking holes in walls. As teenagers, they learned to patch and repair whatever they had damaged before their parents came home from work.

Mark Kelly had joined the U.S. Merchant Marine after high school and studied engineering at their academy. Nearly twenty-one hours after the start of the Gulf War—the first war with Iraq in 1991—Kelly climbed into his A-6E two-man fighter jet on board the USS *Midway* and steered the plane toward an airplane maintenance hangar in northern Iraq, dodging two Russian surface-to-air missiles by rolling his jet and pitching the multimillion-dollar machine skyward. He wound up flying thirty-nine combat missions during Operation Desert Storm.

He went on to become a test pilot before entering the astronaut corps. Although he doesn't really remember some of the earlier Apollo flights, the later missions inspired him.

"It was the whole excitement behind it and the challenge and the opportunity to leave our planet, which always kind of appealed to me," said Kelly.

But, of course, there are dangers. His daughters, eleven and eight years old, were aware of them. The girls had been close to three of the astronauts who died aboard *Columbia,* particularly Dr. Clark.

"That's something I'm going to have to talk to them about," he said before the flight.

Inside a private room, Commander Steve Lindsey pulled on a layer of polypropylene underwear and then blue, high-tech long johns. These have tubes running through them that carry cold water, which keeps the astronauts from sweltering inside the insulated suit.

Lindsey, forty-five, had flown three times before—including a return-to-flight trip with Senator John Glenn. On July 4, 2006, he was the man in charge. While their schedule was mapped out to the minute by NASA, if something went wrong or things changed, the crew would look to Lindsey for leadership.

A native Californian, he grew up in Temple City, playing saxophone in his high school band and wanting to be an engineer like his dad. Lindsey's father had been raised on a farm and went to class in a one-room schoolhouse. The senior Lindsey got drafted and marched off to Korea, but he returned and finished up high school. He used the GI Bill to go to college and get his engineering degree.

"It was really a tough row to hoe, considering his background," Lindsey said. "And so, in that way, watching him live his life and being able to do what he did, I thought, was a pretty significant achievement and I wanted to follow in his footsteps."

But the younger Lindsey was also bitten by the flying bug and looked for a career in which he could do both. He wound up at the Air Force Academy in Colorado. After graduation, he became a test pilot. Then he wanted to reach new heights.

He thought back to July 1969. Apollo 11.

"When I was a kid, probably eight years old, when Neil Armstrong, you know, first walked on the moon. I remember watching it on black and white TV and like every other kid, I said, 'Wow. I want to do that.'"

That gee-whiz feeling crept back, and he applied for the astronaut program.

"I figured the worst they could say was no, and they got my record mixed up with somebody else's and selected me," he joked.

When he wasn't training at Johnson Space Center, he was waterskiing in Houston lakes, riding his motorcycle, and spending time at church with his wife, Diane, and their three children.

He was also on the landing runway at KSC with the families of the *Columbia* astronauts when the shuttle didn't return.

"I was, for that particular flight, the lead family escort," he told reporters. As the countdown clock continued ticking that morning, well past the time the bird should have landed, Lindsey and other NASA officials escorted the families back to the crew quarters to explain what little they knew had happened. Everyone on board was dead. He comforted the grieving group as well as he could.

Later, he had to tell them that the ship had a hole in its wing and caught fire on reentry.

"It was extremely difficult to go through and I hope I never have to do it again."

Discovery's flight that day was the second following the *Columbia* disaster and the 115th for the space shuttle program. It was Colonel Lindsey's fourth flight after earlier missions in 1997, 1998, and 2001.

Spacewalker and U.S. Air Force Lieutenant Colonel Mike Fossum walked out of the private room to sit in a lounge chair as he prepared to put on the launch and landing suit.

Fossum, forty-eight, was the tall, lanky, bespectacled Boy Scout of the group, earning the highest, most difficult rank of Eagle Scout as a teenager and serving as his son's scoutmaster. Every summer, he takes groups of boys on camping trips in the Canadian wilderness and northern New Mexico.

He grew up in McAllen, Texas, and was also inspired by that first moon landing. But he gave up on the idea of being an astronaut, figuring "that wasn't a job that real people had the opportunity to do." His dad worked for the U.S. Department of Agriculture, and he described his parents as people from modest backgrounds.

"They believed in education and hard work and self-reliance," he said about his parents.

He became a proud Aggie at Texas A&M University, where he studied mechanical engineering. When he arrived, the dorms were full and he needed a place to stay, so he joined the school's ROTC program. That accidental enrollment propelled him into his military career. After graduation, he then joined the U.S. Air Force, went to their institute of technology to study systems engineering, and went on to test pilot school. He wound up at Johnson Space Center in flight support.

"I worked down the hall from the crew office and got to know a number of the folks. I realized that they were not that different after all, you know, more or less normal," he said. "It just dawned on me over a period of a couple of years that this was something I could do, too. It meant working a little harder in what I was doing and believing such a thing was possible."

Fossum got the call to be an astronaut on Columbus Day 2003 while he was at home, working on cars and cleaning the garage.

"I fell to my knees and said, 'Thank you, God. I can't believe it's really going to happen.'" He sat there for another twenty minutes in disbelief that his dreams would come true. Then he shared the news with his wife, Melanie, and their four children.

Like Nowak, he was enjoying his first shuttle flight. He quietly didn't like Nowak, though, eventually describing her as "prickly" or "contentious." But he would have to trust her as he dangled from a robotic arm more than 220 miles above the earth, repairing a piece of the International Space Station and performing tests on shuttle tiles to see if they could be repaired in the freezing vacuum of space.

In the suit-up room, the astronauts were helped in putting on the bulky, cumbersome space suit by six "insertion crew members" and fourteen suit technicians from United Space Alliance. The astronauts sat in reclining chairs to go through the hour-long process.

Before the space shuttle *Challenger* explosion on a freezing morning in January 1986, the astronauts had not been required to wear the pressurized suit. But stunned NASA officials realized some astronauts might have survived the freefall into the Atlantic from 48,000 feet up if they had worn them instead of simple blue jumpsuits. Some of them did live for nearly three minutes, until the crew cabin smashed into the ocean and sank to the bottom. Evidence shows that the air packs of Judith Resnik, Ellison Onizuka, and pilot Michael Smith were turned on.

Stephanie Wilson stepped into her orange suit and then scrunched through the black back opening, doubled over. She shoved her arms into the sleeves and wriggled to get her hands to the openings, almost like a toddler getting into a snowsuit. The suit provides full-pressurization and anti–g-force inflation—keeping the blood flowing in the astronaut's body as the spaceship reaches crushing speeds. It is certified to protect an astronaut up to 100,000 feet in altitude.

Wilson, thirty-nine, grew up about an hour outside of Boston in the small town of Pittsfield, Massachusetts. At night, up in the Massachusetts hills, she could look up and see the stars emerge in the heavens. At thirteen, a teacher assigned her class to interview someone in a job the students might be interested in doing one day. She talked to an astronomy professor.

"I was very fascinated by his work," she said. "Later, though, I became more interested in engineering and I thought that aerospace engineering would be a good combination."

She entered Harvard University in the fall of 1984 and graduated four years later with a degree in engineering science. She left for the University of Texas at Austin, where she earned a master's in aerospace engineering. Test pilot school wasn't for her, though. Instead, she went to work for Martin Marietta and then joined the Jet Propulsion Laboratory in California to work on the Galileo project.

NASA wanted to send a spacecraft to Jupiter to study its moons and send a probe into the giant planet's atmosphere. Galileo went up on space shuttle *Atlantis* in October 1989 and reached Jupiter six years later. There were many firsts for Galileo—first to measure Jupiter's atmosphere, first to fly past an asteroid, and first to discover that asteroids have moons. It found evidence of underground saltwater on three Jovian moons—Europa, Ganymede, and Callisto. NASA sent the spacecraft plunging into Jupiter's atmosphere on September 21, 2003, destroying it before it crashed into Europa.

Wilson was the lead CAPCOM for the shuttle *Columbia*'s final mission, talking with the astronauts for days as they worked in space.

She was the second African American female astronaut in space, after Mae Jemison. Wilson is intensely private; in interviews with NASA, she didn't talk about her parents or her home life, only saying she likes to snow-ski and collect stamps. Court records show Wilson had been married and divorced once.

Wilson had to learn to swim in order to be an astronaut, because they spent so much time in the pool at the astronaut training facility, and to be prepared in case of a splashdown during liftoff or landing.

"Between April 1996, when I was selected, and August, when I was supposed to start, I was basically learning to swim," she said. "I was very lucky that the California Institute of Technology coach agreed to teach me to swim. But it was barely enough time. I just was not a strong swimmer. I passed all of the tests, but there was not much margin."

Her task was to work alongside Nowak on the shuttle's robotic arm, and they were dubbed the "Robo Chicks" during training. Wilson's main job on the mission was to use the arm to pick up a giant cargo container from the payload bay and move it over to the space station at an early point in the mission. After that, she served as backup to Nowak.

"It's nice to have another woman on the flight, and so we can help each other out if needed," Wilson told *Florida Today* of Nowak. "But she has always been there if I've had a question or had a need."

Nearby in the suit-up room, spacewalker Piers Sellers shoved his head up through the neck ring's black neoprene dam. Once his head emerged, the dam formed a seal around his neck.

"It's designed to be the width of your neck, not your head, so it's very tight as you push your head through it," explained astronaut Michael Foale in an interview with NASA. "Your hair gets pulled out as the rubber comes down over your face and squeezes your neck."

Sellers, fifty-one, was the witty Brit of the group, always cracking jokes. He was married, with two children. As a boy growing up in Sussex, England, he closely followed the Gemini and Apollo missions.

"Apollo landing on the moon, which just completely captivated me," he said about his inspiration. "And from then on, there was always the hope that I would get to do this."

As a teenager, he started taking flying lessons through the Royal Air Force Cadet Program.

"I was flying aeroplanes around at the government's expense, and having a wonderful time, before anyone would let me drive their car."

He also immersed himself in physics, math, and biology. He got an ecological science degree from the University of Edinburgh and then went on for a doctorate in biometeorology from Leeds University. He studied how Earth's atmosphere and biosphere work together.

"I've always been drawn to aviation and science and this was pretty much the sharp end of both of those things."

Sellers did fieldwork in such far-flung locales as Kansas, Russia, Africa, Canada, and Brazil.

He started with NASA as a research scientist at Goddard Space Flight Center in Greenbelt, Maryland. This was his second spaceflight—his first was just a few months before the *Columbia* accident. Like Fossum, he would be dangling at the end of the shuttle's robotic arm and boom, with Nowak inside the space station controlling its movements. One bad move on her part could kill Sellers. He knew it. She knew it. But he liked and trusted Nowak.

Their German crewmate, Thomas Reiter, stood up in his pumpkin suit, and a technician zipped down the black back zipper, which closed at the base of his spine. Then a white lanyard was attached to it, which went between his legs and hooked onto the suit near his neck. In case of an emergency, he could quickly use the lanyard to unzip the suit.

At his feet, another technician shoved on his black boots, which are made to support the ankles if a parachute jump is necessary, but light enough to allow him to run from a malfunctioning spaceship.

Reiter was using the shuttle as a billion-dollar bus in order to reach the International Space Station. His presence would bring the ISS crew up to three people. The married father of two would also bring some much-needed cheer to the group with his guitar-playing skills.

"It's a great feeling to have a guitar, floating in front of the window, and playing some tunes," he told *Florida Today*.

Like several other astronauts, Neil Armstrong's unparalleled step inspired Reiter's career choice. He watched from his home near Frankfurt.

"When I was 11, I very well remember the first moment when Neil Armstrong put his feet on the moon," he said. "That was actually the point in time when I thought, 'Hey, this would be a good profession.' I thought, 'This must be magnificent if you could stand on the surface of another planet or the moon.'"

Even though neither the Germans nor the rest of the Europeans had a space program at the time, Reiter, forty-eight, went on to study aero-space technology. He also went into the military and learned to fly.

He spent his evenings explaining to his wife and children what he did that day to bolster their confidence and lessen their fears.

"They understand that all the people that are involved in preparing such a mission are making really all efforts to make it as secure as possible."

Finally, Nowak put on her gloves and helmet and turned them into their locking rings, making sure they fit correctly. Her official NASA portrait shows her in her pumpkin suit, her cheeks round and rosy, her long, brown hair brushed and shiny, her blue eyes sparkling, a bright, warm smile spread across her face, and a smattering of freckles peppering her nose. On the ring finger of her left hand was her wedding set—a solitaire diamond engagement ring nestled next to a gold band. She looked younger than forty-three years old and still retained a glimpse of the girl-next-door look of her high school years. There was a minimum amount of makeup—eyeshadow, mascara, and lip gloss. It was the picture of a woman who seemed warm, confident, and welcoming.

Nowak was six years old when Armstrong and Buzz Aldrin walked on the moon.

"I was fascinated," she said. "I wasn't aware at that time that women weren't involved in that kind of thing, so it wasn't something I thought I couldn't do."

Later on, she said, in junior high and high school, she participated in programs to introduce girls to the space program.

But it wasn't until she was in test pilot school that she thought being an astronaut might actually be feasible. In 1996, there were 2,500 applicants, 120 of whom were chosen to go to Johnson Space Center for a week of interviews and a rigorous physical exam. Out of that, 35 were selected for the class of 1996—and Nowak was one of them.

In a preflight interview with CBS News, Nowak said she was more worried about driving a car on a highway than flying on board *Discovery*.

"Obviously, something could happen," Nowak said. "The risks are there, but we've done so much to minimize everything that we can, I feel really confident we're going up with the safest vehicle that we can and that we have a plan in place if something does come off. I feel really good about it."

Technicians removed the gloves and helmet until Nowak and her crewmates were on board the shuttle.

They also stuffed each of the suit's pockets with various items Nowak would need in case of an emergency: flares and a radio, a notebook and pen.

But the most precious item she was bringing with her belonged to her grandmother.

"My grandmother died a few years ago at the age of almost 100, and one of the special presents that she left me was her very beautiful engagement ring, and I'll be taking that with me," Nowak told *Florida Today*. "It's probably one of the most treasured items that's going up there."

She also took with her a small wooden statue of an owl, the mascot of Luxmanor Elementary School in Rockville, Maryland, where she had attended school.

After flights were scrubbed on July 1 and 2, Nowak's husband and children spent July 3 with her parents, two sisters, and their four children at SeaWorld, keeping in touch with regular phone calls to Nowak.

"We decided we had seven kids, and we're all waiting for the announcement, whether they're going to scrub the flight or what," Nowak's father told a *Washington Post* reporter as they waited in line for the Shamu show. The week before, almost two hundred friends and family held a reception for Nowak to wish her well on her flight. Her father said everyone from Nowak's nursery school days to the U.S. Naval Academy attended the bon voyage party.

Now, they were back at the space center.

As Nowak suited up, her high school classmates were gathered in a Maryland park for their twenty-fifth reunion, watching the launch preparations on a television.

"Everybody's talking about it—everybody's excited about her doing that," classmate Dennis Alloy told the *Washington Post*. "It's like 'We know an astronaut!' She's probably the most famous one we have in our class."

Looking over the room on launch day, the seven astronauts were ready. At nearly 11:00 A.M., it was time to head down the hallway, get into the elevator, and ride down to ground level, where cheering space workers and photographers were waiting for them. American flags were handed out, although Reiter proudly waved a small German flag. Armed guards dressed in fatigues and bullet-proof vests, carrying automatic weapons, escorted the waving, smiling astronauts to the van and then followed behind them in an armored car.

The crew loaded up into a silver Airstream Astrovan, plugging into cooling units for the twenty-five-minute ride out to the launchpad. An "insertion technician" (a NASA term for someone who helps the astronauts into their seats on board the shuttle) rode with them, while other technicians waited for them in a special area just outside the shuttle door called the White Room.

Chief astronaut Kent Rominger and several VIPs were also in the Astrovan with the astronauts. As it made its way along State Road 3, a black NASA helicopter hovered overhead, with gunmen seated in the open doors to protect the precious cargo. It followed as the van turned right onto the road leading past the Vehicle Assembly Building, the press site, and the Launch Control Center, which includes the famous firing room with the huge windows overlooking the launchpads. That's where launch control workers sit to monitor via computers the shuttle, the external fuel tank, the solid rocket boosters, and the astronauts.

The van stopped in front of Launch Control to let out Rominger and the VIPs. Media from all over the world left the large press room and individual press buildings to gather outside in the parking lot, on the roofs of their cinder-block offices, and near the road to wave to the astronauts.

After covering the space program for more than twenty-five years, *Florida Today* space reporter Todd Halvorson had several superstitious traditions on launch day. Number one: no donuts allowed in the *Florida Today* block house—a two-room, bare-bones structure—because delays have coincidentally happened on the days when someone brought in the treats. He says, "Donuts denote delay."

Number two: Todd always had to wave at the astronauts and say, "Come back safely," as they drove away from the press site, even though the astronauts couldn't hear him. Someone else had to add, "Even if it's today." Finally, Todd brought in a chicken wishbone and broke it at launch time.

While there are traditions, it doesn't compare to what the cosmonauts do before getting into their vehicle—each one urinates on a tire of the van because that's what Yuri Gagarin did before he launched in 1961. Female astronauts and cosmonauts pour a cup of their urine on the tire.

On July 1, during the crew's first attempt to launch, the crew stopped again, unloading by a deep ditch near the shuttle, which was making creaking sounds due to the freezing liquid hydrogen inside the external fuel tank. The cold audibly hammered into Florida's blistering July temperatures.

Astronauts were allowed to say a final good-bye to their families near the launchpad. The astronauts park along a roadway, while families gather on the other side of a ditch. It's informally called "the wave across the ditch." On the other side of the ditch were the astronauts' families, including Nowak's husband, Richard, their fourteen-year-old son and twin four-year-old girls.

Nowak told the *Catholic Monthly* that she waved to her family and one of the twins began running toward her—stopped by a security guard.

"That was hard," she said.

Astronaut Pam Melroy told *Florida Today* in 2001 that, as they drive toward the launchpad, everyone starts to fade away.

"You start to realize that everybody is turning back except you," Melroy said. "You're the only crazy one headed straight for this giant ticking bomb out on the pad, and it's just you and your crew, and you all kind of look at each other and think, you know, this is just us now."

At about 11:30 A.M., three hours before launch time, the seven-member crew arrived at launchpad 39-B. They walked around the entire pad, the shuttle and gantry looming above them, to take pictures of each other and, as usual, make jokes. As the machine hissed and groaned above them, they received a "go" from Mission Control to board.

Locals and Kennedy Space Center workers know that a colony of turkey vultures lives at the space center. A whole flock of them can usually been seen at the Saturn/Apollo complex, a football-field-sized building with a relic of an Apollo Saturn V rocket hung from the ceiling and a small moon rock that visitors can touch. Following a near-miss of a vulture during the previous launch, NASA was taking special precautions to keep the vultures from flying over the shuttle during launch. Workers removed roadkill and other dead animals from areas around the launch site and sounded an alarm before launch.

The astronauts took an elevator nearly twenty stories up and, one by one, followed a painted path known as "The Yellow Brick Road" as they entered the White Room. It is a sterilized environment, with seven workers wearing white suits and white caps who finish readying the astronauts. Each one wears a different, large number, too, to allow for unmistakable identification. There were five United Space Alliance employees, along with a fellow astronaut not assigned to the mission, and a NASA quality inspector.

One by one, the crew each made a pit stop in a bathroom, wriggling their butts out of the back of their pumpkin suits and the layer underneath. Back in the White Room, each astronaut used a finger to write their initials in the accumulated frost of the eighteen-inch liquid oxygen supply line near the bathroom, a long-standing astronaut tradition.

Commander Lindsey waited first to board the orbiter. The closeout crew slipped the straps of a heavy, cumbersome backpack with a parachute onto his arms and inserted orange glow sticks into his shoulder pockets. Then, they put the black "Snoopy" cap over his head that contained his communication equipment, including ear speakers and a microphone.

They placed Lindsey into his seat—looking out the flight deck windows, his was the left front seat. They left him to lie on his back for several hours as the other astronauts went through the White Room and were placed into their seats; then all seven waited for liftoff.

Rominger said it is not unusual for astronauts to fall asleep during this waiting time. Lindsey, however, had a checklist of things to do.

"Your back starts to hurt because you've been lying on it a long time in this really uncomfortable suit," astronaut Melroy said.

Reiter was next in, heading down to his seat on the mid-deck.

Pilot Mark Kelly went next into the right seat beside Lindsey, followed by Stephanie Wilson, who went down below to the mid-deck.

Mike Fossum and Piers Sellers mugged for the ever-present cameras, with Fossum grabbing Sellers's face and smooshing it into a smile. Sellers waved and said, "Hi, Mom!"

Fossum was seated in the right "passenger side" rear seat in the flight deck, while Sellers was sent to the mid-deck below, where he would fly with Wilson and Reiter.

Finally, it was Nowak's turn—the moment for which she had waited and trained for ten years. It was a moment many of the astronauts and space workers thought might never come after the *Columbia* disaster three years earlier. And many members of Congress continued to question whether billions of dollars in funding for the space program was worth it.

Nowak smiled and laughed with the White Room crew—the picture of composure and professionalism. She was last to climb into her seat, next to Fossum and behind Lindsey. The closeout crew then locked on her gloves and helmet.

At nearly half past noon, the closeout crew sealed *Discovery*'s hatch, waited for pressurization to reach capacity, and left the astronauts behind as the group of seven waited for launch. The White Crew took the elevator down and left the launchpad.

"Makes me proud to be an American," Rene Arriens, forty-five, a White Crew team member, told *Florida Today*. "It never gets old. It makes me well up every time."

Things were very busy outside the orbiter.

The orange external fuel tank had been filled up and made ready to launch in the early-morning hours. Workers filled it with 500,000 gallons of supercooled liquid hydrogen, along with a liquid oxygen oxidizer. It supplied fuel to the space shuttle's three main engines. With the liquid supercooled to −423 degrees Fahrenheit, the external fuel tank developed ice on its outer surface—even in a Florida July's 90-plus degrees. Thus, a group known as "the Ice Team" did its final inspection several hours after fill-up, looking for any problematic chunks of ice on the external fuel tank that could impact launch safety. According to NASA documents, the Ice Team is composed of seven NASA and contractor employees, who use binoculars and a telescope to focus on hard-to-see areas. It was a briefcase-sized ice chunk that knocked a hole in the forward section of *Columbia*'s left wing, leading to the horrible disintegration upon reentry that destroyed the ship and killed the crew. No one wanted that to happen ever again. The Ice Team's inspection took about three hours.

Engine cut-off tests were performed throughout the morning. Anything wrong with those sensors and all bets would be off for a launch.

Astronaut Mike Bloomfield flew the Shuttle Training Aircraft to monitor landing weather during the launch countdown. There can be no lightning-producing clouds within ten nautical miles of the launchpad and flight path, and twenty nautical miles of the landing strip, within thirty minutes prior to launch time. In addition, there could be no thick cloud coverage over the pad and landing site in case the pilot needed to return the orbiter to the landing strip at the Space Center. Weather also had to be nearly picture perfect at the other landing sites around the world.

Inside the fabled firing room, with its rows of computers and tall windows overlooking the launchpads, weather analysts were looking at radar images to make sure no storms were headed toward the launchpad. Engineers in Houston were doing communications checks with all seven astronauts. And the U.S. Coast Guard, Navy, and Air Force were patrolling the skies and waters of the launch zone.

Just before 2:00 P.M., Mission Management Team Chairman John Shannon, NASA Test Director Jeff Spaulding, and Launch Director Mike Leinbach each polled their team members for "Go" or "No go" for launch, with everyone responding, "Go!"

"OK, team . . . *Discovery*'s ready, the weather's beautiful, America's ready to return the space shuttle to flight, so good luck and Godspeed, *Discovery*," Leinbach told the crew.

"I can't think of a better place to be on the Fourth of July," Lindsey said, adding that the crew was ready to give Florida "an up-close and personal look at the rockets' red glare."

With nine minutes before liftoff, the Ground Launch Sequencer computer, located in the firing room, began controlling the countdown automatically. It monitored more than one thousand critical functions through liftoff.

A vulture flew past the launchpad. After space shuttle *Discovery* hit one of the birds in July 2005, NASA installed an antivulture device, which emits loud sounds to scare the birds away from the launchpad.

The launchpad was cleared of all personnel. At T-Minus seven minutes, the Orbiter Access Arm, which included the White Room and access to the shuttle, was retracted, but it could be put back into place in less than thirty seconds in case of an emergency.

At T-Minus five minutes, steam could be seen coming from *Discovery*'s three main engines. Miles O'Brien, then CNN's space reporter, explained it was the boil-off of liquid hydrogen.

At T-Minus four minutes, Lindsey began gimbaling—or moving— the three main engines at the bottom of the shuttle one at a time, back and forth, to make sure they were working properly. He did the same with the shuttle's tail.

At about T-minus two minutes and thirty seconds, the astronauts were told to close and lock their visors and to start their oxygen flow. Nowak's clear visor was already down, but she pulled down the separate, darkened visor, pulled a metal ring down to chin level, and then lifted the darkened visor back up.

The cap on top of the large, orange main fuel tank was moved off.

At T-minus 90 seconds, pilot Mark Kelly was fidgeting, rocking back and forth with his arms up, and pushing in between each of the fingers of his gloves.

At T-minus 60 seconds, Nowak sat patiently, holding a shuttle manual on her left leg and a notebook and pen on her right leg, repeatedly clicking the pen in her right hand.

At T-minus 31 seconds, pilot Mark Kelly flipped three switches inside *Discovery*'s cockpit to start each of the three auxiliary power units. Then the orbiter's onboard computers took over for launch.

At T-minus 10 seconds, flares were ignited under the shuttle's three main engines to burn away any residual hydrogen gas that may have collected.

At T-minus 6.6 seconds, the main engines began firing, and Nowak and her crewmates began shaking in their seats.

At T-minus 6 seconds, the water sound-suppression system started flooding the flame trench underneath the vehicle with 300,000 gallons of water—not to put out fires, but to protect the ship from the tremendous energy of sound waves rushing away and up from the engines and solid rocket boosters.

At T-minus 0, the solid rocket boosters ignited and the bolts that held the shuttle to the launchpad were severed, giving the machine 3 million pounds of thrust from each solid rocket booster.

"And 3 . . . 2 . . . 1 . . . and liftoff of the space shuttle *Discovery* returning to the Space Station, paving the way for future missions and beyond!" NASA public information officer Bruce Buckingham said at 2:38 P.M. on the only Fourth of July NASA launch in the space program's history.

A mile from the launchpad, a medical team waited in a bunker, ready to rush to help the astronauts in the event of an emergency.

About three miles away at the press site, with the wind blowing toward him from the launch, Halvorson could feel the sound waves rushing through him, shaking internal organs and raising goosebumps from head to toe.

Astronaut Jim Reilly told CNN's Miles O'Brien that the noise at the press site is far more impressive than it is on board the shuttle.

The light from the main engines and booster rockets' fire, even in daylight, is nearly blinding.

Along the beaches north and south of Cape Canaveral, along riverfront parks, and in boats in the Indian and Banana Rivers, hundreds of thousands of people watched the most technologically advanced machine ever built ride atop a column of white exhaust and blinding flames to reach the heavens. Cheers and screams rose up, tears welled in people's eyes, and parents explained to small children what was happening. Then the sound waves rumbled down the beaches and rivers.

Inside the firing room, NASA Deputy Shuttle Program Manager John Shannon had tears in his eyes. "It just blew me away," he told *Florida Today*.

Those who remembered 1986 held their breaths for more than a minute, waiting for the command of "Go at throttle up," the point at which the *Challenger* exploded after a leaky O-ring in one of the solid rocket boosters allowed fuel to escape, catch fire, and then explode. After that point, unenlightened onlookers felt the astronauts would be safe. But those who knew better relaxed after more than eight minutes, when the shuttle reached low Earth orbit.

Inside *Discovery*, an excited smile spread across Nowak's thin face as the rocket began to shake and then lift off. The astronauts heard a loud rumble and the force of the launch pushed them back into their seats. One astronaut described it as a sharp jerk—like being in a car accident. It knocked the breath out of them as the shuttle shook violently and shot upward.

Astronaut Scott Kelly told his brother Mark that, at launch, "You'll feel like maybe something is going very wrong. You'll feel every pound of thrust. It's full power instantaneously."

As always, it was a bumpy ride as the engines and rockets roared and howled under them. *Discovery* cleared the launchpad, pushed into the afternoon sky, and then, several hundred feet up and ten seconds into launch, rolled onto its back for more efficient use of fuel.

"There's a little more elbow room than in the Apollo capsules," astronaut Buzz Aldrin told me in 2007, comparing his experience in Apollo 11's small space capsule to the space shuttle. One of the first two men on the moon also said that the Apollo rides were a little smoother, since they sat on the top of the rocket, farther away from the flames and engines.

Houston took over shuttle operations after *Discovery* cleared the tower.

The streak of sunlight from the window above Nowak slid across her closed visor. As a mission specialist 2, she was serving as the flight engineer, "who is as familiar as the pilot and commander about the shuttle systems and how they work," she said.

Mark Kelly appeared to have his right hand over his heart as they launched, while Fossum switched his helmet's visor to the right to look out the window and see the coastline of Florida falling away below them at 3,000 miles per hour.

"Sitting on 4½ million pounds of explosives—it's not inherently safe," Fossum had told reporters a few months before liftoff.

At 60 seconds, they were traveling at supersonic speed. At more than 70 seconds, CAPCOM Steve Frick gave the command for "Discovery, Houston. Go at throttle up."

"Roger. Go at throttle up," Commander Lindsey answered.

The powerful g-forces pushed on the astronauts' chests. It became hard for them to talk, and they breathed in small pants.

At two minutes, the solid rocket boosters—the white rockets on either side of the orange external fuel tank—expended all of their fuel of Ammonium Perchlorate Composite Propellant and dropped off, smoothing out the ride for the second stage of the launch.

On Florida's east coast, *Discovery* looked like a blinding star getting smaller in the sky.

As they raced for orbit, sophisticated radar and more than one hundred high-resolution cameras tracked every moment of the launch. Engineers tracking the flight saw small foam fragments flying off the shuttle's redesigned external orange tank, but NASA officials said *Discovery* was traveling too high and too fast for it to do any damage.

If the first stage of launch is a roller-coaster ride, astronauts say the second stage is like being shot out of a cannon, a very smooth constant acceleration.

Outside *Discovery*'s windows, the blue of Earth's sky turned into the blackness of space.

When they made it into orbit at fifty miles above Earth, Nowak and her crewmates lifted their helmet visors. At 3½ minutes into the flight, the shuttle was fifty-three miles high, traveling at "a clip of more than 4,000 miles an hour."

Fossum extended his left hand, and Nowak grabbed it with her right as they shook hands. And at that point, NASA officials congratulated the newbies on the crew—they were officially astronauts because they were in orbit.

At four minutes and eight seconds into the flight, they were sixty-one miles up, traveling 5,000 miles per hour, and past the point of no return to Kennedy Space Center. Thirty seconds later, they were midway to the International Space Station.

At six minutes, *Discovery* rolled onto its stomach.

At about eight and a half minutes into flight, as the shuttle continued to climb "uphill" through Earth's atmosphere, glowing orange plasma flashed around the shuttle, and Lindsey cut off the shuttle's main engines.

Nine minutes after liftoff and seconds after the external fuel tank separated from the space shuttle, Fossum and Wilson used handheld cameras to videotape and take still photos of the orange tank to see if any large portions were missing.

Outside the windows, Earth's sphere dangled in intense blue, white, beige, and green Technicolor splendor, more clear, more vivid than anything Nowak had ever seen before.

It was the greatest journey of Nowak's life. Although it was not to be the most famous.

12

★　★

Is That Your Final Answer?

Detective Becton knew that he needed to find Nowak's car. He knew something inside it would give him some of the answers he needed—like what was her real intention?

"So, where's your car?"

She mumbled something.

"Do not, please do not make me track down and run every license plate and every vehicle in this airport because that is gonna be just about as much disrespect any person could show me. Now where's your car?"

She sat silently.

"You're not protecting anybody by [not] telling me where your car is."

"Okay," she said.

"Where's your car?"

Again, she said nothing.

He knew she didn't want anyone outside of his office to know what was going on. He knew she could lose her job—and probably would—when this became public.

"I tell you what—we can do this one of two ways. You're gonna either tell me where your car is or I'm gonna, I'm gonna go call KSC security and I'll find out where your car is, and I'll tell them why I'm looking for your car. And, and I'm just going to stop this whole thing because you know what, I'm tired of getting jerked around right now. You're wasting my time and you're insulting my intelligence."

"I . . ."

He cut her off. "I appreciate the fact that you want to protect people, but right now, you're not in a position to decide . . ."

"I . . ."

He cut her off again. " . . . if you're able to protect anybody."

"If you would tell me," Nowak started to say.

He stopped her a third time. "You know what. You have a bachelor's in science, in aerospace engineering in the Naval Academy, a master's of science in aeronautical engineering, and a degree of aeronautical and astronomical energy engineering from the U.S. Naval Postgraduate School," he said, reeling off a list of her accomplishments from her NASA bio. "I think you're intelligent enough to realize what is going to be the best situation. And I think you're intelligent enough to realize that you're insulting my intelligence."

Becton knew that many times women would rather curl up and die than offend anyone. Becton was using this tactic against her. And he was trying to play good cop/bad cop—with only one officer.

"I'm trying to help you control damage as much as possible. You're not giving me anything to help you out with. And all I've gotten to ask you, 'Where did you park your car?' In . . ."

This time, she cut him off. "How will that help?"

"Well, right now, we don't have any trust between you and I. I have you personally half screwed-up in the head right now, okay. And I'll try and get help for you."

She mumbled something.

"First, you have to earn my trust, okay?" he said. "I might be able to help you control some of the damage that's been done. I don't know because I don't know what's going on in your head, okay."

She thought about that for a second. She knew things were bad.

"How did you know she was going to be parked in the Blue Lot?"

"I didn't."

Becton hadn't realized yet that Nowak had stalked Shipman since Shipman's flight had landed at around midnight.

"Okay, so out of one, two, three, four. . . . six parking lots in a 20-square-mile area, you just got lucky and picked the Blue Lot?" he asked.

"I only knew about the Blue and Red Lots at the airport," she said.

"Okay, so you took a guess . . . Where's your car?"

"I promise, I will tell you. It's not like I have a choice anyway, you know that," Nowak said.

"Well, actually, you do have a choice. You can not tell me and then I can go through hours of looking . . ."

"Wait, wait! I know," she said.

" . . . through a list and find it."

"I know. I know," she said again.

"So, you do have a choice."

"You're going to the, you're going to go to the car," she said.

"Okay."

"I'm gonna tell you where the car is," she said.

"So far, you haven't done anything to earn my trust in that area. And let me explain this to you—honesty is everything with me," he said. "Grant it, you haven't lied; you haven't said much, though. Now, I'm not here to [give] out information. I'm not a public information officer."

From what I've gotten out of this interview, Lisa was trying to kill Colleen, he thought. *I need a lot more to prove this beyond a reasonable doubt, though, and hopefully that "a lot more" is sitting in her car.*

"Are you going to tow it?" she asked.

"Well, if it's towed, I can guarantee you that it's not gonna get broken into first. We have a lot of auto burglaries out here."

"Okay."

"Second, the next thing I'm gonna ask you, if I have consent to search your car, in which case, I'll need the keys to your car."

"What if I say no?" Nowak asked.

"Then I'll get a search warrant for your car and I'll search your car anyways."

"No, I'm not going to do that. I just want, I, I still don't understand what's going to happen to me . . . It sounds like this is the only place you're going to talk with me, then you're done with me and I, I, just . . ."

Nowak didn't watch a lot of television.

"I, I know I put myself in this situation," she said. "I just don't understand, you told me where I'm going."

Becton turned his head a little and squinted his eyes. "Huh?"

"I just don't understand . . . what happens after that."

"Okay. I kind of figured this is a quid pro quo, so I've already answered part of your question," he said. "I'm waiting for the answer for parts of my question. So where's your car?"

"I feel like [it's] the only thing I value, I have left to give you, and if you cannot help me anymore . . ."

"Okay . . . let me explain it this way.

"I . . ."

"Listen now, don't start thinking like you know what I'm gonna say."

"Okay," she said.

"I'm not the type of man that is going to tell you something and then not follow through. I'm definitely not like the man that is just gonna string you along for a ride. Now the thing is this, I'm not gonna lie to you to control this interview, either," he said, lying. Which the U.S. Supreme Court has ruled is legal for law enforcement officers.

"Okay, fine," Nowak said. "It's not about control."

"It *is* about control," Becton answered. "You've been controlling a lot because you have, and here's the thing about this type of individual I see you as—tell me if I'm wrong—you are very particular about what you say and how you say it. You deal mostly with logic. If it doesn't make logic—if you can't make logic of it, it doesn't count. And the problem is, you can't make logic out of your actions, either. Okay. Now you're very much a thinker . . . Why don't you want to tell me where the car is?"

She is trying her hardest to control this interview, he thought.

"I know. I'm gonna tell you," Nowak said. "I'm just trying to think if there's anything that I need to tell you before you get there. I don't want, I don't want any surprises."

"Okay, first tell me where the car is."

"Okay, I know you said that you're a man of your word. I know there was some discussion about her, about November. You didn't tell me that."

"Okay, she told me they met in November. Um, pretty much, that was it. She told me she knew of you, and you guys worked together. That was all between you and him. She did not mention anything about a relationship. She did not make any reference about a relationship," Becton said, lying again.

"She didn't mention a relationship?" Nowak asked.

"No. Actually, I talked to him without her in the room."

"He's here?"

"No. He's pretty much told me the same things you have, except he had a, he's made it more like it's a work relationship than anything," Becton said.

"So, he doesn't want to get me into any more trouble." She knew she could be court-martialed for conduct unbecoming of an officer for having an affair while she was still married.

"Well, I'm going to indicate that both she and he indicated it was a work relationship only. If you're telling me . . ."

"It appears to me that that's what she believes," Nowak said.

"Mmm, hummm . . . and if you're telling me that to protect his career, then it's fact, that's something I can, I can state. And according to the two of them, it's a fact," Becton said. "You don't believe you were spending too much time on a personal level with him, so now I've got three people saying it was just a work relationship, although you wanted it to be more."

Officer Timothy Ryan stepped into the room, looked at her and then looked at Becton. "In the Blue Lot, nothing appears."

"The time is right now," he told Nowak.

"Okay, I am. Let me make clear I understand what they are saying about wanting more . . ."

She was like a dog with a bone. She was so focused on the relationship that she didn't understand she was about to be charged with several felonies. But Becton didn't tell her that, either.

"I have some emails that, that disagree with that," she said.

"From his computer?" Becton asked.

"Yes."

"Did you have his password?"

"Yes."

"Okay, that's how you knew."

"No, that's not how I knew, I, I—it was not anything different."

"You need the emails verifying he was telling the truth?"

"Yes—it was nothing more different than what he had told you. There is no indication of what I was trying to find out about the other stuff . . . and . . . the phrase 'just a work relationship.' I'm sure you can understand at this point that this is something you say if everyone else in the office is a . . ."

"I believe what you tell me," Becton said. "He has every reason to tell me lies—you don't."

"It's important, whether it's me being this deranged person that is over-interpreting things, or that my . . . ," Nowak said.

"I can validate the assessment in the fact that, all right, it fits as to what happened tonight. Um, to this point. And the reason I can say that is because you don't really have anything to lose. He and she have a lot to lose."

"Like what?" she asked.

"Well, she's a commander, I believe, or he's a commander? They're both in the military, how many officials are in the military? That, in and of itself, is something at risk for them, okay. You don't benefit by telling me anything that's not the truth right now."

I can talk in my own circles, too, Becton thought.

"I guess I'm trying to figure out if there is any way that I can minimize the impact to all three of us," she said.

"I'm working on that. I understand that this is very delicate in nature based upon the position that the three of you hold, okay. And I can tell you that, that it's not my position as a human being to embarrass anybody, or allow anybody to be embarrassed."

"I also, it's important. I'll be at . . . after going off on this that I completely misinterpreted what wasn't there and what was never there and what was not said to me," Nowak said. "But I don't want to cause the trouble that I am now in that I don't want it to make my case first worse either."

"I understand."

"Well he's not being accurate that he told you that."

"And that's gonna reflect upon him . . . not through anything you and I have done per se, okay. Let's hold on a second. I'm running out of time," Becton said, completely making up pressure from his supervisor in order to pressure her to give up the car's location. "My boss is ready to walk in here and ask me where this car is, okay. I've got to provide him with some answer. Are you not gonna tell me where you parked and then we can continue?"

"Okay, do I need an attorney?"

"That's the decision you have to make. You are an educated woman, very intelligent woman. I cannot give you any advice."

Becton glanced at her military awards: Defense Meritorious Service Medal, given for noncombat outstanding achievement; Navy Commendation Medal, given for heroism or meritorious achievement or service; Navy Achievement Medal, given for commendable performance in routine duties or exceptional achievements that have not otherwise been recognized. She had also earned the NASA Space Flight Medal.

But two awards stood out: Navy Rifle Sharpshooter Ribbon and Navy Sharpshooter Pistol Ribbon.

"What would you do?" she asked him.

"Um, I have to, I will get in trouble if I do give advice, okay," he said. "I can't even switch shoes, okay."

"Okay, um . . ."

"I can tell you that I see some issues with you personally that I think can be taken care of, that will help make things better for you. And I will be more than happy to recommend to the state attorney's office and when you get an attorney, as well," Becton said.

This woman desperately needs a psychiatrist, he thought.

"Because I really think that that will make all the difference in the world, in a positive way," Becton said. "But I need to know where the car is right now."

"So, I'm going to have a trial?"

"That's up to you. You have many, many options . . ."

Becton went on to describe to her that she would have what's known as a "first appearance" before a judge, where she would hear the charges against her and enter a plea of guilty, no contest, or not guilty. And she could get an attorney.

Inside the tiny interrogation room, what neither she nor Becton knew was that her family was already working on hiring an attorney. She also didn't realize that she was facing a possible prison term.

"Um, there is possible some other outcome?" she asked.

"Yeah, it's possible. Now where's your car?" he asked for about the dozenth time. He would ask that question more than sixty times before the night was over.

"Want something to eat?" Becton asked Nowak.

"No, thank you," she said. Her stomach was in knots and the thought of eating made her a little queasy.

"Can I get you some crackers or pretzels or something? You want to try that?"

She mumbled something.

"There is no way not to call security?" she asked about NASA security.

"I'll look into [it to] see if I don't have to notify security. I can't make you any promises. Eventually, they will find out, though. They do continual criminal background checks on you, am I correct?"

"I don't know."

"But then the question comes up: Is it worse to live and find out, or is it best to tell them now?"

"I wouldn't . . . I have to put it down . . . I'm just thinking about . . . ," she said, her voice trailing off.

"I will look into trying to avoid doing that. That's the best I can do right now," Becton said.

"Who has already been notified after you and the two of them?"

"The two of them [have] been notified. My chain of command has been notified and the FBI has been notified," Becton said.

"The FBI? For what?"

"You're part of a crime that involves two federal employees. I haven't given any names. I just told them that there are two NASA employees [who] were in the office and I was interviewing them over an incident. So they don't know who it is right now. Not to mention, he is in Miami, working the Super Bowl."

I think the seriousness of the situation is beginning to sink in, he thought.

"But don't . . . if they're wondering details, don't you . . . ?"

"Knowing my buddy, he's just leaving it up to me to work with. This way, I can say I notified the FBI and everything is okay. It's more of a CYA for me than anything else. There's no official notification or anything."

"Why did you have to notify them, then?"

"Well, one is this is federal property. If I don't notify, if there's an incident of this magnitude on federal property, I'm gonna get fired. So, my buddy and I have a working relationship. We understand each other's thingamabob and we work within those parameters, okay?"

"Is there any possible way I could go back to work tomorrow?"

Oh, no . . . she still doesn't get it.

"Don't worry about it," he said, trying to distract her from the tsunami of reality that he hoped would wash over her soon.

It's like she lived in some kind of fantasy world.

"I'm not gonna prevent you from going back to work. I don't see any reason why you can't go back to work. I just, from what you've told me, repercussions will be severe, knowing NASA. And you might get kicked out of the military."

She began to whisper something that Becton couldn't understand.

"You're a captain in the military. And trust me, as many captains that get in trouble and [they're] still captains."

"Not like this."

So she did understand.

"You'd be surprised. You're not the only captain that has ever been in trouble like this. In the history of the Navy, I promise you, you're on a very long list of captains. Trust me—I grew up in the Marine Corps, I've seen . . . I've seen a lot."

Actually, there weren't too many Naval captains whose escapades made international headlines, as hers would in just a few short hours.

She asked to write something down on a piece of paper, scribbling it in tiny letters.

"Can I see where the car is?"

"I wrote it down. You will get it either way. I just want an answer."

Becton wasn't sure anymore who was controlling the interview.

"Okay, this isn't gonna be *Let's Make a Deal*, okay. You're gonna let me know where the car is, and when I get a call, I'll let you know what legal says. Or you're not and I'm gonna go find it."

"There's stuff that's in there, um, that you know I'm going through the process of getting divorce paperwork, and examples that were provided to me are in the car. And you can guess who they were from."

"Um, do I have your permission to go in your car and search your car?"

Should I get a search warrant? he thought. *If this were an average case, I'd just normally get consent first and, if that failed, get a warrant.*

"I'll bring you with me so that, if you change your mind, you're able to make sure that I didn't change your mind."

"Are you changing what you're accusing me of?"

"No, I'm still charging you with aggravated stalking," Becton said.

"That's it?" she asked.

"Battery and theft is a lesser crime for the pepper spray."

"What's aggravated stalking?"

"That's a felony, whereas the battery is a misdemeanor. That gives the assistant state attorney bargaining room if they want to negotiate a plea deal."

"But it's the same . . ."

"Sort of, kind of," Becton said. "The stalking is everything that led up to tonight . . . [T]he battery is the actual pepper spray. So it's not a double jeopardy, but it gives the assistant state attorney some room to plea things down, if that's what's agreed upon by everybody."

13

★ ★

Walking on Sunshine

Discovery was racing toward the space station at 17,500 miles per hour that July 4, with Nowak and her six crewmates on board.

The STS-121 mission was to do a detailed inspection of the shuttle's heat shield, test new safety and heat shield tile repair techniques, as well as deliver supplies, equipment, and European Space Agency astronaut Thomas Reiter to the International Space Station (ISS). *Discovery's* crew of seven astronauts and the space station's two men would work in tandem to achieve each of the mission's goals, a highly skilled team trained in the complex orchestration of tasks that would take place each day. Nowak was a crucial member of this team and its mission.

"The striking thing about this shuttle flight is that the most important task is simply survival of both the vehicle and the astronauts who are riding in it," the *New York Times* wrote in a July 5 editorial.

The space shuttle fleet cost billions of dollars to assemble, but, since no additional spacecraft would be built, the three remaining—*Discovery, Atlantis,* and *Enterprise*—were simply priceless. They were the most advanced mechanical devices humans have ever built, designed to lift off, fly in space, make the delicate maneuvers to dock with the space station, and then return to Earth unscathed. They were also, in essence, a fleet of cargo-carriers, moving parts, supplies, and crew to and from the space station. And the astronauts were the smartest handymen anywhere in the solar system, required to maintain and fix anything on board the space station—inside and out. They also move cargo into the space station and garbage out of it, conduct advanced scientific experiments, and represent NASA in the best light possible during interviews before, during, and after their visit to space.

Discovery carried more than two tons of cargo, including a long, round specialized cargo container named Leonardo, on its fourth flight.

Carefully packed in the months and weeks leading up to the launch, this container was hauling:

- a French-built "Minus Eighty Degree Laboratory Freezer" to store experiments on board the ISS. The freezer was also designed to transport samples to and from the ISS in a temperature-controlled environment;
- a European gas-tight incubator for biological experiments;
- a new oxygen generation system. This device was considered "an equipment test," with potential for use on any future long-duration flights to the moon and Mars. It was also designed to help the ISS support a crew of six in the future;
- a new bicycling machine for the ISS crew—working out is critical to astronauts because they lose bone mass during spaceflights. Workouts keep them in shape and help to stave off boredom; and
- a replacement air conditioner for the ISS.

If NASA had known just how serious the damage was to *Columbia* in 2003, a rescue mission could have been sent to the ISS to pick up *Columbia*'s crew. Thus, NASA created detailed inspection plans for all space shuttles after launch, and each ship's thermal protection systems—hundreds of black tiles, each one handmade out of a material called "reinforced carbon-carbon" on the shuttle's nose and wings' leading edges, along with black ceramic tiles on the shuttle's underbelly. The tiles were designed to withstand the superheated temperatures of reentry and protect the ship's thinly coated skin, but then cool down after landing.

If the comprehensive inspections of the shuttle *Discovery*'s heat shield, conducted once the vehicle was in orbit, had suggested that it would not be able to survive a reentry, or another problem occurred, then NASA's plan was for space shuttle *Atlantis*, commanded by Brent Jett, to mount a rescue mission. The shuttle crew would wait on the ISS for rescue.

Discovery STS-121 was the first shuttle mission to carry a cable designed to connect the upper flight deck manual controls, used during landing, to an aviation electronics unit in the mid-deck, allowing mission controllers in Houston to land the shuttle unmanned if required. However the *Rescue Flight Resource Book* stated that a controlled breakup—rather than a landing—was planned. They would ditch the multibillion-dollar machine in the ocean if a major problem arose.

After their busy launch day, most of the crew zipped themselves into sleeping bags just past 7:30 P.M. Houston time, attached to the walls of the crew cabins, and slept.

Day 2–Wednesday, July 5, 2006

It is a long-standing tradition for crew members' family, friends, and even the ground crew to choose a song for each astronaut to awaken to. It gives the crew something to look forward to and is seen as a morale booster.

At about 4:00 A.M., Eastern Standard Time, on Day 2, the New Galveston Chorale's version of "Lift Every Voice and Sing" began playing for Stephanie Wilson.

"That one is particularly near and dear to my heart because it's performed by a local group in the Galveston area," said Wilson, who NASA described as a rookie enjoying her first morning in space. "Particularly after the day of our nation's independence, it's very fitting because it reminds us that anyone and everyone can participate in the space program. Thanks very much and good morning."

Nowak wasn't feeling well. Like about half of the astronauts who fly into space, she was experiencing space sickness as her neurovestibular system adjusted to weightlessness. In other words, she was throwing up.

"The first day, the first 24 hours, I didn't feel so great and it took me a while to adapt," Nowak told CBS News. "Then I started feeling better."

After breakfast and preparing for the day, Commander Steve Lindsey opened the large cargo bay doors that run the upper length of the shuttle's body. Nowak then used *Discovery*'s $600 million robotic arm to grasp and extend a fifty-foot-long Orbiter Boom Sensor System (OBSS)—NASA-speak for a long, jointed pole, tipped with two types of lasers and a high-resolution television camera. Nowak used the OBSS to inspect the underside of the shuttle for damage. Particular attention was paid to the leading edges of the shuttle's wings. The inspection was a 6½-hour, highly focused, and intense task.

Just after 6:00 A.M. Central Time, Nowak stood in front of a rear-facing window, with two closed-circuit monitors to her right, all of which gave her a view of the shuttle's fifty-foot robotic arm, with the boom now attached to it. A camera for one monitor was mounted to the forward part of the cargo bay, while the second camera was at the rear of the bay. Just outside the window, which faced the cargo bay, was the robotic arm emblazoned in red with the name "Canadarm" for the country that built it. Nowak used a joystick in her right hand, along with a round, rotational hand controller in her left, to manipulate the robotic arm and boom to stretch out over the shuttle's wings, crew cabin, and nose. She would capture about twenty minutes of data each for both wings and the nose cap, with the remaining 5½ hours used to maneuver the delicate arm and boom.

As a fixed camera peered out over the boom, a line of hexagon sun glares flared on its lens.

"This is Mission Control Houston with live video from the camera of space shuttle *Discovery* as Mike Fossum, Lisa Nowak, and Stephanie Wilson conduct a survey of the crew cabin of the shuttle using this robotic arm as a unique vantage point to take a look at the condition of the thermal protection system blankets that cover the crew cabin section of the nose of *Discovery*," an announcer's voice intoned as the images were shown on live television. "Again, all this is part of the varied inspections going on today to verify the health of *Discovery* so that it can return home to Earth at the conclusion of this mission without issue as it passes through the earth's atmosphere. The crew has completed scans using a laser dynamic range imaging system on the end of a fifty-foot-long orbital boom sensor system of the leading edges of both of the space shuttle's wings, as well as the nose cap, where the sensitive reinforced carbon-carbon material protects those hottest areas of the shuttle from the 3,000 degrees of reentry temperatures."

"*Discovery* crew, per the captain's survey, you have the go at thrust-on at point 1:11," Steve Lindsey said.

"Copy and recorder is on," Nowak answered him.

"Thank you," Lindsey said.

"Notified the crew is commencing the official survey of the crew cabin using the camera on the robotic arm," Houston narrated from the ground as crews in the United States watched closely.

"We're looking all over the leading edges and the underside—all the places that we might have damage," Nowak said of the task. Her job that day was one of the most important of the entire mission. "We also did some looking with [a] camera—you can see there's just some of the jets and you also see some of the tiles . . . All this went down [to analysts on the ground]. Imagery team did a great job analyzing it and sending us data and letting us know our spaceship looked really clean."

The inspections had found a minor issue—a gap filler between two tiles was protruding on the port side lower wing, not a location of particular concern; and it was noted that it had been there since the shuttle was constructed in 1982.

The inspections also revealed the presence of bird droppings on the leading edge of the right wing. The launch director noted that he saw droppings in that location before launch. The droppings remained there through landing—though rather charred.

"The bottom line is we do a 100 percent inspection of the heat shield—because we don't think any of our ground-based sensors or any of the cameras are good enough to positively confirm we haven't had a problem," NASA shuttle program manager Wayne Hale told *Florida Today*. "We base our reentry on clearing that heat shield by the inspection we do."

Following lunch, Nowak and Wilson worked in tandem to return the Canadarm and the Orbiter Boom back to their berth on the starboard sill of the cargo bay. Then they and Fossum spent an hour using cameras to photograph the exterior of *Discovery*'s crew cabin, which is not protected by the black tiles or reinforced carbon-carbon tiles. Wilson also used a hand-held camera to shoot pictures of the base of the shuttle's tail fin.

Wilson and Reiter also readied items to be transferred to the ISS.

As they all worked on photographs, spacewalkers Fossum and Sellers, helped by Kelly, checked out and prepared the spacesuits they would use for their spacewalks in the coming days.

Day 3–Thursday, July 6, 2006

On their third day in orbit, at about 2:30 A.M., Elton John's "Daniel" drifted into *Discovery*, played for German astronaut Thomas Reiter, from his sons Daniel and Sebastian, along with his wife, Consuela.

It served as "a reminder that there are people on Earth thinking and waiting for us who give us the strength to do what we're doing," Reiter said.

That day, *Discovery*'s crew prepared for the International Space Station two-man crew—one Russian and one American—to inspect the shuttle's underbelly and then dock with the space station.

The ISS operates in low Earth orbit, about 220 miles up—nearly the distance between Washington, D.C., and New York City. It was constructed through a partnership between the United States, Russia, Japan, Canada, and the European Space Agency, with six crew members usually on board. It is the largest man-made object in the sky and can often be seen as a bright light, dashing across a velvet landscape of stars. Viewing it from the ground, it appears different from an airplane in that its light does not blink . . . just a silent sentinel of space exploration slipping by overhead.

As *Discovery* approached the ISS, Lindsey commanded a pitch maneuver prior to docking, showing the shuttle's underbelly to the ISS crew so they could briefly inspect and photograph the heat shield.

"Just a beautiful sight to see," NASA lead flight director Tony Ceccacci said from Houston. Many NASA workers, along with employees of the various contractors that work on the shuttles, view the machines in a very personal way—almost as though they were living, breathing creatures. Many feel a genuine affection for the spacecraft.

The ISS was flying about fifty yards above *Discovery,* as the Pyrenees Mountains bordering Spain and France passed 220 miles below. Lindsey guided the shuttle through the slow-motion back flip.

"Eventually we flip all the way over—and now our nose is about pointed toward the ground," Kelly said. "When we saw the space station coming over tail we were all pretty happy about that—'cause you like to see it show up where you expect."

When *Discovery* was flat on its back under the ISS, Nowak commanded: "Start photo. Start photo."

Standing at windows in the ISS's Russian command-and-control center, outpost commander Pavel Vinogradov and flight engineer Jeffrey Williams snapped 350 high-resolution digital photos of the black tiles lining *Discovery's* belly.

"All this data will be combined and reviewed by several different separate teams of imagery analysts on the ground to make sure that no stone is left unturned as they verify the health of the shuttle's reentry protection systems," Houston announced.

Nowak then commanded, "End photos."

The critical window for photography lasted one minute and thirty-three seconds.

Using the shuttle's thrusters, Lindsey inched the spaceship into place, finishing his final approach at a "glacial speed" of one-tenth of a foot per second, *Florida Today* reported. The shuttle's round hatch is situated in the cargo bay, and Lindsey had to lift the shuttle up to the ISS airlock using a computer, video monitors, and his own line of sight out the shuttle window to navigate. The rendezvous technique was developed by Apollo 11 moonwalker Buzz Aldrin, when he obtained his doctorate of science from the Massachusetts Institute of Technology in 1963. It remains in use today.

Docking rings on both craft touched, and NASA-designed latches "grasped both ships in a metallic embrace," gingerly docking the shuttle with the ISS, *Florida Today* reported. It was just before 10:00 A.M., Houston time, as the pair coupled above the South Pacific's Pitcairn Island. They pressurized the vacuum between the two vehicles, pumping air into the lock. Ninety minutes later, the astronauts on board the ISS welcomed the new company from *Discovery*, helping them to open the hatch. Lindsey pulled himself into the ISS's U.S. Laboratory, performed a weightless somersault, and gave a thumbs-up. Reiter officially became a member of the ISS Expedition 13 crew, bringing the ISS team up to three for the first time since the *Columbia* accident. It was the first time ever that the ISS crew consisted of an American, a Russian, and a European.

"Congratulations on this first full house," Canadian Space Agency astronaut Julie Payette said from NASA's Mission Control Center.

"Yes, it's a full house," Williams replied. "The climate here has changed significantly."

The first order of business was a mandatory safety briefing inside the ISS, followed by the installation of Reiter's customized Soyuz seat liner into the Russian rescue vehicle. They also checked his pressurized Sokol flight suit. Then the astronauts began the task of unloading five thousand pounds of food, clothes, equipment, experiments, and spacesuits. In zero gravity, even Nowak, the most diminutive of the crew, could carry the heaviest box by merely pushing with one hand—or even one finger.

"You pack up all your belongings in a box and you throw them on a truck. Basically, you bring them to your new house and you pile them in a room, and then you take some time and put it in the right rooms and the right order," *Florida Today* reported that flight director Ceccacci said. "That's basically what the crews will be doing."

There was also cargo packed up and tied down in "hallway" areas that would be returned to Earth nearly two weeks later.

About three hours after docking, the shuttle and ISS crews prepared Canadarm2—the robotic arm on the space station—from inside the ISS's Destiny Lab. As clearance restraints around the shuttle's docking mechanism do not allow the shuttle arm to grapple the boom on its own, they used the ISS robotic arm to lift the boom from *Discovery's* payload bay sill, then hand it over to the shuttle arm, operated by Lindsey and Fossum. It was an extremely delicate maneuver.

Following dinner, the crew began to sleep in shifts.

Day 4–Friday, July 7, 2006

At 2:15 A.M. on July 7—flight day 4—it was Nowak's turn to have a special song played for her. Her children chose the Beatles' "Good Day Sunshine."

Nowak responded that they "have a sunrise every 90 minutes and have plenty to laugh about. It's certainly something to talk about from my first, first spaceflight. Thanks everybody. Good morning!"

While most men in space don't have to worry about their short hair getting in the way or creating a corona around them, Nowak opted out of having "big space hair." "I've seen pictures of some people that come back and got their big-hair picture, zero-gravity hair, but I'm just going to have mine in a ponytail and keep it under wraps and hopefully out of my way," Nowak told *Florida Today* with a laugh.

But she was more concerned with doing the perfect job at running the robotic arm, knowing she had her crewmates' lives in her hands.

"The biggest thing we have to watch out for there is not getting bored or complacent," Nowak said in an interview. It's "kind of like driving a car if you're tired, it could be easy to let something happen if you're not watching very carefully."

Nowak, who was notorious among her peers at both the Naval Academy and NASA for getting lost, told CBS News that she was a little disoriented moving through the ISS.

"After we got to [the] space station and you have to turn different directions to enter each module, that took a while to get used to," she said. "But I'm happy to say I think I'm completely adapted now."

That day, they were busy. Nowak and Wilson got back on the joystick, using Canadarm2 to move the cargo container Leonardo, with thousands of pounds of equipment, from the shuttle's payload bay, lifting it in place to be attached to the ISS via another airlock. Wilson was the lead on this operation, moving it over beginning at 4:45 A.M., with the assistance of Fossum, Sellers, and Kelly.

Due to concerns that loose straps on the cover of a latching mechanism might interfere with a proper seal, Lindsey and Reiter conducted leak checks of Leonardo. By 7:15 A.M., the seal was found to be safe, and they entered the cargo container. Wilson took command of the transfer activities.

Using the Orbiter Boom, Nowak and Wilson conducted six additional focused inspections of *Discovery's* heat shield tiles, fewer than had been required on the previous mission. They targeted areas on the nose cap, as well as two gap fillers protruding from *Discovery's* belly. They also wanted to take a closer look at a piece of fabric near the shuttle's nose. In all, they found six areas of interest.

Nowak knew the risks and the dangers of her job, thinking back to when she was a Navy test pilot.

"One time, we were a couple of hundred miles out over the water and we lost a hydraulics system—and you know, at any moment, if something else fails, there's all kinds of consequences," Nowak told *Florida Today*. "So there's been a number of challenging incidents that I think being in the Navy has prepared me well for this job."

The robotic arm was developed when NASA realized they would need a way to move giant, heavy pieces of equipment from the shuttle's cargo bay and to the ISS's airlock systems. According to a NASA report, "it didn't take them long to settle on the idea of a glorified crane, something that could pick the equipment up, move it over and let it go."

On the space shuttle fleet's twelfth flight—*Challenger* in June 1983—ice built up on a vent on the bottom of the shuttle, causing a potential danger for reentry if it had broken off and damaged the shuttle. The crew needed a way to get it off, and used "a little tap from Canadarm."

The robotic Canadarm, which has evolved over three decades, is capable of both heavy lifting and an incredibly delicate touch.

Canadarm has rescued satellites and was invaluable in helping astronauts repair the $2.5 billion Hubble Space Telescope, which was launched into low Earth orbit in 1990 with a faulty main mirror, compromising its ability to photograph images correctly. Using Canadarm, astronauts "grabbed" the telescope, pulled it into the cargo bay, and performed some minor surgery, fixing the mirror. Hubble has undergone four other servicings and upgrades.

In 2001, Canadarm2 was added to the International Space Station. It can reach farther, carry more weight, and maneuver into more positions.

The shuttle's Canadarm and the station's Canadarm2 "worked in tandem on almost every shuttle flight to build the station, handing off school bus–sized components between them," according to NASA.

As cargo transfers and more safety snapshots were taking place, Fossum and Sellers prepared for their spacewalk on Saturday—two days away. They organized the tools they would need, along with readying the Quest Airlock. Although they had prepared and practiced repeatedly on Earth, they still reviewed spacewalk procedures.

Lindsey and Kelly talked with radio reporters from CBS, FOX, ABC, and National Public Radio. The ISS crew also spoke with CNN, CBS, and the Associated Press.

Because everything was going as scheduled, Mission Control decided to add an additional day and a third space walk to the mission. They had trained for it on Earth and were expecting it. The shuttle astronauts went to bed by 6:00 P.M. that Friday evening.

Day 5–Saturday, July 8, 2006

On flight day 5, the stereo from Mission Control began playing the strumming of a guitar just after 2:00 A.M. on board *Discovery.*

"Lord of all creation, of water, Earth and sky, the heavens are your tabernacle and glory to the Lord on high. God of wonders beyond or galaxy, you are holy, holy . . ."

Fossum's family chose the song for his first-ever spacewalk—or what NASA calls an EVA, for Extra Vehicular Activity.

"We wish you a glorious day as you execute a flawless EVA and marvel at the wonders of the universe," Houston read a message from Fossum's family.

"I want to thank my wonderful wife, Melanie, and kids for selecting one our favorite songs this morning," said Fossum, who wore a silver cross on a chain for the flight. "I do think it's particularly appropriate as Piers and I prepare to step outside for about four and a half trips around this chunk of creation we call Earth. It is going to be a great day! Thanks a bunch and good morning to you."

Fossum and Sellers prepared for the mission's first spacewalk. They first did what's called a "pre-breathe" exercise, working out on stationary bikes to purge their systems of nitrogen as they breathed in pure oxygen for a total of two hours and twenty minutes.

Then crew members helped them put on their cumbersome white spacesuits, boots, gloves, and helmets—looking very much like the Apollo astronaut spacesuits that were used to walk on the moon. They could tell the two men apart because Sellers's suit had red stripes on the arms and legs. He served as the lead spacewalker, having two EVAs under his belt from October 2002.

Once the men were sealed into their white spacesuits, the suits filled with oxygen to pressurize them, and the men breathed in pure oxygen, again to wash their bloodstream of any excess nitrogen. Without these preparations, any nitrogen in their blood could lead to tiny gas bubbles in their bodies and pain in their joints, giving them what's known as "the bends," because of the excruciating pain that occurs where their bodies bend. Scuba divers are also susceptible to this condition.

"Astronaut Stephanie Wilson has been helping with the preparations for this spacewalk, now seen in this live view from the International Space Station Quest Airlock, with astronauts Mike Fossum and Piers Sellers in their spacesuits," Houston narrated in a live broadcast with a familiar, calm voice, as Wilson helped them maneuver. "The two men are scheduled to step outside of the Quest Airlock at 8:13 A.M. Central Time this morning. They're now in the midst of an in-suit, pre-breathe procedure, which puts them breathing pure oxygen for 60 minutes."

As the men breathed the pure oxygen, they also put on a jet-propulsive backpack that would help them return to the spacecraft if, for some reason, they became untethered from *Discovery*. During EVAs astronauts are double-hooked to either their shuttle or the International Space Station at all times. The standard procedure is to wear that backpack as a triple backup.

The pair exited the Quest Airlock. Both Fossum and Sellers were tethered to the boom at the sill of the cargo bay, with tool boxes and tools attached to their spacesuits. They were floating over the Strait of Gibraltar on a southeasterly course across the Mediterranean.

They got into position and installed a portable foot restraint for

Sellers to ride on the end of the boom sensor system, so he could evaluate its ability to support astronauts repairing the thermal tiles, should the need ever arise.

"Ok, um, Lisa, we'll need to . . . to the tether anchor position," Sellers said to Nowak, breaking up slightly.

Sellers slid into place, one hundred feet of the Canadarm and boom extended toward the ISS. First Sellers did what the men affectionately called "bouncing on the boom," simulating what it would be like to do repairs on the shuttle in space.

"Ok, Piers, I imagine you're ready?" Mark Kelly asked.

"I'm ready," Sellers answered.

"Ok, Piers, first test point, you're going to retrieve the digital camera, you're going to pick a target, take two photos of the same target and then stow the camera [away]. Two, one—go for the camera."

As a cloudy sky over Spain slipped beneath him, Sellers grabbed the camera with his bulky gloves.

"I've got the camera in my hand," he announced. "Taking photos of other guys—one . . . two . . . put my lens cover back on."

The camera was tethered to Sellers, and he simply let it go when he was finished.

Kelly continued to watch from within the ISS.

"Ok, Mike, do you, if you see any motion, if you could just quantify it, please?" Kelly asked Fossum.

Fossum began laughing. "Yeah, tough to tell motion against the moving background."

Kelly agreed and said he saw almost no motion.

"Next point is a slow layback," Kelly said. "Three . . . two . . . one."

"There's a pretty view," Sellers said as he slowly bent back in a cosmic limbo, his feet lodged in the foot restraint platform. "I'm coming forward again."

Nowak moved the boom back into the cargo bay to pick up Fossum. But Fossum suddenly had a problem with his eighty-five-foot-long safety tether, which he had accidentally left in the locked position. It was damaged as a result.

"Oh no! . . . That's embarrassing," he said, realizing his mistake. The damaged tether had to be switched for a spare.

Fossum then joined Sellers on the platform and attempted increasingly vigorous operations that simulate various potential repair scenarios.

"Copy. . . . both . . . coming out of the bay," Nowak announced as she slowly moved them out of the shuttle's cargo bay, dangling them in space over Earth.

Chris Hadfield, chief of robotics for the Astronaut Office, told *USA Today* that Nowak and Wilson's work with the robotic arms was "as complex a robotic . . . ballet as we've ever done in history."

Nowak had to make sure to maneuver the arm and her crewmates around or over ISS hardware, antennas, and the shuttle without damaging anything—or injuring Sellers or Fossum. There were close calls before Nowak's flight. In 2001, the arm almost collided with an antenna, and, a year later, it nearly rammed the shuttle's door in the cargo bay, *USA Today* reported.

So, one wrong move on Nowak's part could have bumped them against the ISS or shuttle, damaging equipment or, in a worst-case scenario, rupturing their spacesuits, which could have killed them almost instantly.

"Roger that," Sellers acknowledged.

Both Fossum and Sellers said the boom stabilized quickly from their movements, making it a good work platform.

"Hey, Piers, take a second and look at the Earth here," Kelly radioed from the flight deck. "I think you've got Ireland and England coming there."

"Oh!" shouted Sellers, who was born in Sussex. "Oh my goodness, it's a beautiful day in Ireland!"

The ISS, with *Discovery* mated to it, and two astronauts flying out a few yards beyond, slipped over Europe and moved 220 miles above the Caspian Sea.

"I'm in a dream. Nobody wake me up," Fossum said, looking down.

A few minutes later, the sun had set again, and Earth was dark below them, with only webs of electric lights splashing across continents.

Once they were finished, Nowak took them on a "long, slow, dark ride back toward the payload bay," Fossum said. "And it does get cold out there, especially when it's night and you cannot see anything made by man—a little lonely, too."

After seven and a half hours in the inhospitable environment of space, Fossum and Sellers headed back to the airlock, making their way hand over hand toward the ISS's laboratory module and the Quest Airlock.

"Just give me a second to make the jump over here," Fossum said.

"Yep. Once you get oriented, you go first, right?" Sellers answered. "Looks like sunrise."

Fossum looked toward Earth. "Here comes the sun."

Sellers advised him to keep his visor up, and then the pair climbed into the hatch, ending their spacewalk at 3:48 P.M. The pair proved that, in the event of an emergency, astronauts would be able to use the Canadarm and the boom to perform repairs on a space shuttle.

NASA had a request for the entire crew that day, sounding like the parent of a toddler:

"Just as an example of how closely Big Brother watches, we would like to remind you, remaining fluid in drink bags should not be poured down the [Waste Collection System, also known as a toilet], but stowed in Wet Trash—yes, they can tell."

NASA was concerned that additives in the drinks might mix with a chemical in the toilets, which could cause a clog.

As the spacewalk was being conducted, ISS Commander Pavel Vinogradov unloaded more cargo.

Day 6–Sunday, July 9, 2006

On day 6, the crew was nudged from their sleep at 2:08 A.M. by ABBA's "I Have a Dream," chosen by Mark Kelly's two daughters, Claudia and Claire.

"Good morning, Houston!" Kelly said. "I want to send a special thanks to Claudia and Claire. Sometimes it's like being in a dream up here. Floating is a big part of that. The view outside the window is incredible right now. I'm not sure what we're over, but we can see some land down there. I really want to send a special thanks to them and I miss them."

The ISS crew was awakened half an hour later with their normal alarm.

One of the day's highlights was a press conference from orbit. All nine NASA crew members participated, beginning just before 11:00 A.M. One of the questions came from Keith Cowing of the blog *NASA Watch*, asking how they can convey the importance and the substance of what the astronauts do to keep the public's attention. After some nervous laughter, Commander Lindsey answered.

"I came into this program ten years ago. And I came into it because I believed in it—and [I have] always believed that, as a country and as a world, we have invested in our future. One of the ways we invest in our future is to do research and by pushing technology," Lindsey said. "[I]t's not easy. And if you ask anybody here they'll tell you that it is not easy to do. But you know, when you do something really hard and you have to push technology and come up with new ways to do things . . . you discover new things and you come out with new spin-offs that help everybody here on Earth. So, I think research—just like they do in universities—is the same thing. By doing things that are hard, that is how we advance."

And he added: "There have been huge benefits from the space program in the past and there will be in the future. And that's why I am here. And I believe it, and I think if you ask anyone here, they will tell you the same thing."

There were more cargo transfers, with Wilson taking charge of the operation again. She ultimately oversaw the transfer of more than 7,400 pounds of equipment and supplies from the Leonardo and another 1,800 pounds from the shuttle's mid-deck. That included a spacewalk suit and emergency jetpack. As Leonardo was emptied, the crew began filling it with 4,400 pounds of material no longer needed on the space station, including experiment supplies, trash, and broken equipment.

Fossum and Sellers remained busy preparing for the second spacewalk, organizing their tools and readying their spacesuits. Then the pair, along with Lindsey, Kelly, Wilson, and Nowak, went over EVA procedures.

Deputy Shuttle Program Manager John Dale also told them what they and every NASA worker on the ground wanted to hear: *Discovery* was cleared for landing, telling the crew all photos and videos showed a safe vehicle.

Just after 5:00 P.M., *Discovery*'s crew went to sleep.

Down below in Houston, the astronauts' families stepped outside of their homes to look up at the dusky sky just before 9:00 P.M. They watched the bright, unblinking light of the ISS and shuttle—the brightest object in the sky—slip past them overhead.

Day 7–Monday, July 10, 2006

Day 7 began a little after 1:00 A.M. with the familiar and repetitive piano chords and drums of Coldplay's "Clocks," played on the stereo, with lead singer Chris Martin singing, "Lights go out and I can't be saved, tides that I tried to swim against . . ."

"That song was from Mandy and the kids, and they hope you enjoy your EVA today," Houston told Piers Sellers.

"Good morning, Houston!" Piers said. "Just got to the mike in time. Thanks very much. That was beautiful. Hello, Mandy, kids."

Sellers and Fossum were set to restore the International Space Station's malfunctioning Mobile Transporter railcar to full operation by replacing a tube of cables supplying power, data, and video to the transporter as it moves along the rail on the outside of the ISS. They were also delivering a spare pump for the station's cooling system. Their work that day would allow for the resumption of construction work on the ISS, which was about half-finished in July 2006.

"This is a live view of the Quest Airlock with the hatch coming open and the thermal cover open," Houston said. "Crew members that, when they put their suits on battery power, it will signal the official start of today's spacewalk."

"Yeah, OK, the hatch is locked up and now I'm going to put the handle in the ready position," Sellers said, flipping open the cloth hatch. "Looks like a country down there. Beautiful sun shining up the river. That's gorgeous."

The spacewalk began at a quarter past 7:00 A.M., Central Time. The astronauts climbed out of the airlock as the station and the docked shuttle *Discovery* dashed over a nighttime Spain, with lights glowing yellow along its coast and in its cities, 220 miles below.

"Nothing to be afraid of in the dark, there," joked Mark Kelly, who was directing the spacewalk from inside *Discovery*.

Florida Today reporter Todd Halvorson wrote that Fossum reconstructed a tagline from the 1979 horror movie *Alien*.

"Everyone can hear you scream," Fossum joked.

"I've got time to get outside, put on my alien costume, and wait for Mike to come out," Sellers added to the joviality.

The pair climbed into the shuttle's payload bay area and lifted up the square 1,400-pound pump. Inside the ISS, Nowak and Wilson manned the joysticks and gingerly grabbed the pump with the station's Canadarm2 to swing it into place. The astronauts then climbed up to the space station's truss area and mounted the pump on a stowage platform. The spare would be used if the pump now on the station radiators were to fail.

Both astronauts soaked in the view.

"The whole world goes pink and then copper and then white in just a few seconds when you're outside," Sellers said of sunrise. "It's really quite the most beautiful thing."

The world revolved underneath them, the continents visible under clouds.

"Mark, what river are we flying over?" Sellers asked.

"Amazon," Kelly replied.

"You're kidding," Fossum said.

"That's a BIG river," Kelly said.

"It is huge," Sellers agreed.

Next they focused their attention on repairing the station's Motorized Railcar, which had broken down the previous December. Shaped like a baby grand piano, the 330-pound assembly motor is equipped with two ribbon-like cables that relay power, data, and video to the railcar. The cables are designed to unreel behind or wind up in front of the railcar as it moves along the station's central truss. Using designed-for-space, pistol-grip bolt drivers, the spacewalkers switched out the crippled part from the metallic backbone of the ISS, replacing it with an identical new one.

"If we don't get this thing fixed, we can't move this truck that moves up and down the front face of the station and we can't continue with assembly [of the ISS]," Sellers said in an interview before the mission. "So we absolutely have to get it fixed before the next mission."

According to USA Today, "Nowak called it 'the most challenging robotics test . . . on this flight' because she'll have only a grainy and incomplete view of what the arm is doing."

With Nowak driving the robotic arm and Wilson backing her up, Fossum carried the damaged assembly to the shuttle's payload bay, riding the station's robotic arm, and handed it off to Sellers.

"Time for an elevator ride down to the payload bay," Fossum told the two women.

He handed off the old assembly to Sellers, took the new one, and rode back up to the truss with it. After dropping off Fossum, Nowak gently moved the robotic arm back to the payload bay and picked up Sellers. She moved him up to the work area, where Sellers joined Fossum and they installed the assembly.

"Don't want to bump anything expensive," Sellers said.

"It's all expensive," Fossum replied.

Twice during the spacewalk, Fossum had to readjust a loose connection on the emergency jet thruster backpack on Sellers's spacesuit. As the pair completed their tasks, Sellers looked back on the day's work.

"I can't believe it," he said. "I mean, really, who makes this stuff up?"

"That was a day," Fossum agreed after the six-hour, forty-seven-minute spacewalk.

Back inside the ISS at 2:00 P.M., the pair tended to their suits. All of the suits' power packs needed to be recharged or changed out, and their air and water tanks had to be replenished.

As space enthusiasts watched the thrill of the spacewalk, Vinogradov and ISS flight engineer Jeff Williams, along with Reiter, worked unceremoniously throughout the day to transfer trash and unneeded equipment and supplies to the Leonardo.

Day 8—Tuesday, July 11, 2006

The familiar, upbeat strains of Smash Mouth's "All Star" blared on the stereo for Nowak, from her family on flight day 8.

"Thanks a lot and good morning!" Nowak said. "It is really cool being a little bit closer to the skies up here. We're having a great time. Thanks!"

The crew gathered in the ISS biology lab to do some tricks with water—which made for amusing shots of Sellers trying to catch water bubbles in his mouth as his crewmates pushed them away from him.

"The cool thing is you can drink the water out of the air," Sellers said, as he snapped his mouth at a half dozen bubbles around him. "End up in insane rodent mode."

Nowak, Fossum, Wilson, and Lindsey also participated.

"Lisa has a very dignified, well-regulated approach to this particular trick," Sellers said as a smiling Nowak caught the bubbles in her mouth. "She takes her time, she chases the last one, she gets it all sorted out."

Nowak, Wilson, and the other astronauts continued to transfer cargo between Leonardo and the ISS. They unloaded new items and stowed 4,300 pounds of science experiment results, unneeded items, and trash.

Among the items ferried to the ISS were a new window and window seals for a microgravity science glove box, used for experiments involving fluids, flames, particles, or fumes. Reiter installed the new window on day 8.

At about 7:20 A.M., the *Discovery* crew talked with the Associated Press and *USA Today*.

Nowak was asked about the final inspection of the shuttle's heat shield.

"We hopefully won't see anything different," Nowak said. "The lighting can change—it depends on what kind of light and darkness you have on each inspection point. That might make it a little bit tougher at some points of the inspection."

The reporter asked Nowak if she thought, as a pilot who had flown many aircraft, that the shuttle could be called operational again after a safe launch, reentry, and landing.

"The shuttle, in a way, is always a test vehicle," Nowak answered. "Every flight is a test flight. But for the way we've been using it in the past and calling it operational then—yes . . . After this flight, I think it will be ready to go."

And a little after 10:30 A.M. Eastern Time, President George W. Bush had a private phone conversation with the crew, telling the astronauts that they "represent the best of service and exploration," and thanked them for the job they were doing.

For dinner that night, the entire crew gathered in the "Russian Bar and Grille" module, as Sellers called it, which has a table big enough to accommodate everyone. Shared meals are, according to behavioral health and performance researchers, a boost to morale and a way to foster camaraderie among crewmates.

The menu included cheese, cranberry sauce, beef with vegetables, lamb with vegetables, pork with vegetables, pike, perch, and "some really, really good mashed potatoes," Kelly said. "They're great."

Dessert was chocolate from the shuttle orbiter's galley.

"This is really good. All of it is," Kelly said. "I mean personally . . . coming over from the orbiter after, you know, a few days of orbiter food, it's really nice to have some of the Russian food. It's really good."

Sellers also enjoyed the home-cooked meal.

"It's very nice . . . it's very homely," Seller said, trying the Russian lamb, potatoes, and onions.

"Tuesday nights is karaoke night down there," he added, to the laughter of his crewmates.

Day 9—Wednesday, July 12, 2006

A student choir began flight day 9, singing "I Believe I Can Fly" for Stephanie Wilson.

On tap that day: a third spacewalk by Sellers and Fossum to take infrared photos of the shuttle wings' leading edge and to try out potential shuttle repair techniques. The men donned their spacesuits once again and, at 6:15 A.M., Central Time, they floated into the space shuttle payload bay.

Nowak and Wilson ferried Sellers, with his white spacesuit's red stripes, out on the robotic arm so he could peer at the wings with the special camera. NASA was trying to figure out if the camera's infrared images would help engineers assess the wings' viability. When he was finished, Nowak and Wilson slowly moved him into the payload cargo bay, near Fossum.

"Hey, Mike. You're doing great, doing great," Sellers joked as he moved into the bay.

Their task was to perform repairs on "pre-damaged" samples of heat shield materials that were arranged in about a dozen small squares on one block of metal.

The repairs were expected to work best when the warmed heat shield material—which Fossum and Sellers called "the black goo"—was cooling off. So, Mission Control carefully coordinated the test repair actions with exposure to the sun—the only natural heat source in space. The procedures and the adhesive material had been tested in a vacuum on Earth, but the zero-gravity tests were required due to gases generated by the material and the potential for bubbles to form that could weaken the structure of the repair.

With Fossum helping, Sellers used a caulk gun to dispense the adhesive material and then putty knives to spread it around on the samples. During the job, Sellers let go of the putty knife—or spatula—and then realized it had somehow escaped its tether.

"Guys, I've got to tell you, my spatula's escaped," Sellers lamented.

He tried to locate it, with the assistance of Fossum.

"Don't worry, it happens," his crewmate told him.

While it might sound like a small item to lose, the tool cost two thousand dollars and could damage the shuttle or ISS if it slammed into something like an antenna or a solar panel.

Mission controllers spotted it over the port side of the shuttle payload bay and determined that it was not likely to cause any problems. Sellers apologized for letting go of the tool and causing the extra work on the ground associated with calculating whether the loose tool posed a hazard.

The 2½-hour experiment itself went well, and the men climbed back inside the airlock after their final mission spacewalk, which took a little more than seven hours to complete.

Day 10–Thursday, July 13, 2006

On flight day 10, the *Charlie's Angels* theme song was played for the entire crew—a gift from their flight training team on Earth, a small army of people at Johnson Space Center who had helped the astronauts prepare for their flights, going over every procedure, every task, every job repeatedly until it all became second nature for the crew.

"Good morning, *Discovery!*" Mission Control announced. "That music, if you can call it that, is from your training team. We're assuming there is a really good story here and we expect to hear it when you get back. Also, your training team would like to say they had a tremendous amount of fun with you over the last couple of years."

After nine nonstop days of shuttle inspections, spacewalks, and moving cargo, the crew enjoyed a well-deserved "day off," although there were still a few things to be done, including media interviews.

The folks at Johnson Space Center also sent a special haiku for the team in a Mission Summary email:

Day of Freedom Launch
Leonardo and Spacewalks
Crew now Chillaxin'

In addition, they had some tongue-in-cheek suggestions for some playtime, including arm wrestling with the robotic arms, playing catch with the Orbiter Boom with the robotic arms, and Marco Polo—noting that "the whole fish-out-of-water thing would need to be tweaked."

Later in the afternoon, Fossum and Nowak participated in live interviews with MSNBC and FOX News.

"What have you guys been eating up there—is the food really gross or is it edible?" one television anchor asked.

"We have a number of different things to eat," Nowak answered with a smile. "One of the most popular items is shrimp cocktail. We've had some other chicken and rice and vegetables . . . [O]ne of the things I have just about every meal afterwards is some chocolate. Which I brought a large supply of and I think all my crewmates are happy to share in that."

Other things included in Nowak's preordered meals were food that must have reminded her of growing up in an Italian household: cheese tortellini, turkey tetrazzini with vegetable risotto, lasagna with meat, and pasta vegetable parmesan.

When asked if they got to contact their families, Fossum explained that they have email and a version of Skype. Nowak added that it was something she enjoyed.

"It's really super to see them moving around and saying 'hi'—that was a lot of fun and also, we get to do emails just like you do on the ground," Nowak explained, referring to her children in the family chat. "We can do that several times a day and keep in contact with them that way. And we can even send pictures in email. That's very important when you're supporting astronauts up here and families down there. Everybody needs that kind of support."

Of course, there was no mention of her strained marriage or even her husband.

Nowak's email in-box included warm wishes from friends and colleagues, along with one from eKnowledge about her son's SAT-ACT prep classes and another acknowledging that her email about prekindergarten classes for her twin daughters had been received and read.

"Hey, Lisa and Stephanie!" wrote Charles "Casey" Joyce, a colleague at Johnson Space Center. "I'm not sure if you guys get your regular email while you're up there, but I thought I'd give it a try just in case. I wanted to congratulate both of you on an awesome job flying the arm! From a robotics perspective, I don't think things could have gone any smoother and we couldn't be any happier."

Joyce also thanked them for passes to see the launch in person.

"You guys look like you're having a great time and having lots of fun!" he wrote.

Any emails that might have been exchanged between Nowak and Oefelein or Nowak and her husband during her flight were not made public by NASA, citing privacy laws.

Lindsey insisted that the crew take advantage of the best part about being in space on this day off.

"I ordered the crew that they were not allowed to work, and they had to go look around station, tour station, take pictures and just have fun," Lindsey said. "And I think everybody took advantage of it, came off of it refreshed and ready to pick up at the last part of the mission."

They spent a few moments thinking about Earth below, they said in answer to reporters' questions about the conflict in the Middle East.

"We just flew over the Middle East," Sellers said, "and I have to tell you, from up here, it looks peaceful and quiet just like the rest of the planet. And I think all of us are mindful, when we're flying around and around this wonderful Earth, that it's all we have. This is humanity's home, and hopefully, one day, we'll all get along."

Day 11–Friday, July 14, 2006

Trumpets blared "The Texas Aggie War Hymn," by the Fightin' Texas Aggie Band, on flight day 11 from Fossum's wife, Melanie—much to the chagrin of Stephanie Wilson, a graduate of the rival University of Texas.

"Gig 'em Aggies!" Fossum said. "And thanks to my lovely Aggie wife, Melanie, and most of our family and many friends for setting that up today."

Starting around 8:30 A.M., Wilson and Nowak used the ISS robotic arm to complete the transfer of the Leonardo cargo container, packed with more than four thousand pounds of material, from the ISS to the shuttle for its return to Earth.

Wilson and Nowak also used the shuttle's arm and extension boom to inspect the shuttle's port wing for any signs of micro-meteoroid damage while on-orbit.

They showed off some "space art" Sellers had created for them—a drawing of two robotic arms with their nickname, "Robo Chicks," written between them.

Day 12–Saturday, July 15, 2006

Mark Kelly's girlfriend, U.S. Representative Gabrielle Giffords, picked U2's "Beautiful Day" for him for flight day 12.

"I want to say thanks to Gabrielle for the great choice of wake-up music," Kelly said. "My crew members were kind of concerned they might have to listen to another ABBA song. They are really happy about that."

It was the final day of joint operations for the shuttle, and ISS and crew members would spend the day preparing to leave.

"It was a great time, but it was time to go," Nowak said. "We think they were really glad to see us when we got there and then I think they're really glad to see us go. We kept everybody really busy."

At 2:30 A.M., all nine NASA crew members assembled for a farewell ceremony. And then six astronauts swam though the airlock into *Discovery* and closed the hatch behind them. Gingerly, *Discovery* undocked from the ISS just after 5:00 A.M. With the six astronauts now on board— Thomas Reiter stayed behind—the crew left the ISS over the Pacific Ocean, just north of New Zealand. Kelly flew *Discovery* to a point above the station before performing two final separation burns.

Nowak and Wilson used the robotic arm and the Orbiter Boom Sensor System to perform the final inspections of the starboard wing and shuttle nose cap for any possible damage caused by orbital debris while docked with the International Space Station. They also put away cargo and personal items in preparation for the landing, and conducted leak checks.

"You could still look out the window and see the station getting farther and farther away," Nowak said.

That day, NASA was also keeping a close eye on a tiny fuel leak in one of three engines—called an auxiliary power unit—that provide the hydraulic power needed to steer the ship during reentry, deploy its landing gear, and brake after touchdown. Because of NASA's love of redundancy, only one of the three engines is required to be operating in order to have a safe shuttle landing.

The auxiliary power unit is fueled by hydrazine. The fuel tank was leaking six drops per hour, raising the incredibly small chance of creating a fire in the shuttle's rear engine compartment.

According to *Florida Today*, "Deputy Shuttle Program Manager John Shannon said the leak would have to be 100,000 times worse before it could burst into flames. Nonetheless, he said engineers would monitor the situation to make certain it doesn't deteriorate."

The astronauts weren't concerned. "It's pretty minor," Kelly said.

Day 13–Sunday, July 16, 2006

The next morning, a song from the Cure was cranked up before 12:30 A.M., selected for Piers Sellers by his family for the day before landing.

"Beautiful! That was 'Just Like Heaven' by the Cure from my wild coffee-drinking days. The days of my youth," Sellers joked. "That is the one song that the whole family likes. So please thank them very much for playing it for me."

Despite having been a pilot on two previous shuttle flights—and flown on a third—despite having been a test pilot for the U.S. Air Force, despite more than seven thousand hours of flying time in fifty different types of aircraft, and despite having traveled more than twenty million miles in space already, Lindsey, along with Kelly, spent part of the day practicing a simulated landing on a computer program called "Pilot."

"It's just a video game, kids, so that's all I'm doing is a video game," Lindsey joked. "I'm highly paid just to play this game."

As entry flight engineer, Nowak, along with Lindsey and Kelly, did an hour-long flight control system checkout, which allowed managers in Houston to assess the performance of the auxiliary power unit with the leak. They also fired the control system jets.

Other activities included a long series of live press conferences from orbit, and preparations for landing, like stowing items on the mid-deck and fitting the seats in place.

Discovery received a clean bill of health on its final heat shield inspections and was given a "Go for landing," relayed to the crew while they were on their midday meal break.

"I think the vehicle has done great. I've flown this vehicle before, and it did great before, and that's just a tribute to the folks at Kennedy Space Center that take care of the vehicle," Lindsey said.

As part of preparations in Florida, "the crews will be walking down the runway, checking for foreign object debris," NASA spokeswoman Tracy Young said.

The center also was preparing to welcome 350 to 400 guests, she said, mostly invited by the astronauts. For this landing, *Discovery* sported tougher Michelin tires. In addition, it used Global Positioning System navigation information as a backup, with the goal of making it a primary system on future missions.

"We're going to take it out for a spin," said Steve Stich, a NASA manager.

Day 14–Monday, July 17, 2006

At 3:13 A.M. on landing day, "The Astronaut," by Something Corporate, was played for Lindsey from his family in honor of the shuttle's landing.

"Well, thanks!" Lindsey told the CAPCOM. "Looking forward to a good day here. Hopefully, with good weather we will be on the ground in about eight to ten hours."

Over the next five hours, the crew and Mission Control went through a series of checkouts to make sure everything was "go" for landing. Mission Control polled multiple managers monitoring various aspects of the flight on computers, and each replied "Go" when queried about their status.

At 4:30 A.M., the crew closed the large cargo payload doors and began to put on their "pumpkin suits," stowed away gear, burned off fuel and trash, and installed seats.

Lindsey and Kelly checked the shuttle landing system and steering jets. At 7:07 A.M., Lindsey fired *Discovery*'s engines for three minutes for de-orbit, the Indian Ocean glowing blue below them. They started their approach about nine thousand miles from the landing site. At 50,000 feet, Lindsey took over the controls to fly the ship in manually. Everyone on the flight deck was strapped in and ready, with Nowak again behind Lindsey. In addition to looking out the windows, Lindsey and Kelly could see a display on the screen in front of them, showing the shuttle's pitch (the nose's up and down levels), along with altitude and speed.

Lindsey flew *Discovery* over Guatemala and Mexico, then circled over the Gulf of Mexico, as he came out of orbit and dropped in altitude. It was done to put him on target for the landing strip in Florida.

Cloud coverage at Kennedy Space Center caused a change in runways at the Shuttle Landing Facility. Lindsey allowed Kelly to steer the ship for a few moments, all as *Discovery* left double sonic booms in its wake—sounding as if two bowling balls had been dropped on a wooden floor in quick succession.

As the landing strip at Kennedy Space Center came into view, they approached in a steep dive at more than 200 miles an hour. Their goal is to always land it like an airliner. Fifteen seconds before touchdown, the wheels extended from the belly of the ship. Lindsey kept the shuttle lined up with the runway. The back tires touched down first with a plume of smoke as they hit the asphalt. A red and white parachute—or drag chute—deployed at the rear of the spacecraft to slow it down. Then the front tire touched down. They rolled to a stop at 8:15 A.M., Eastern Standard Time, after 202 orbits around Earth and more than 5.3 million miles traveled.

Discovery's successful launch and landing led NASA to resume regular space shuttle missions in the construction of the ISS, and Lisa Nowak played a vital role in that, with her flawless control of the robotic arms to photograph the ship's underbelly, to carry her crewmates Sellers and Fossum on the end of the Orbiter Boom, and to move thousands of pounds of supplies and equipment—more than any other space shuttle mission ever.

14

★ ★

Colleen

Colleen Marie Shipman arrived in the world in January 1977, during a twenty-five-day freezing spell, near Pittsburgh, Pennsylvania. One of two sisters in the Shipman household, she grew up waving her Terrible Towel for the Steelers and sporting her black and gold jerseys for the Penguins and the Pirates.

Her tiny body was perfect for the tumbling, balancing, dancing, and flying of a gymnast, which she took up as a girl and maintained as a practice for many years.

Like many students in Pennsylvania, Shipman studied German, excelling at the subject before graduating from Center Area High School in Monaca in 1995, according to her website with Oefelein, Adventure Write.

School Superintendent Edward Elder, who is related to Shipman's grandmother, described her as "perky, bouncy and compassionate" in a 2007 story for FOX News.

"If you wanted a profile of 'the girl next door,' that would be Colleen," Elder said.

She enrolled at Penn State after high school graduation and decided to indulge her two passions: German and chemical engineering.

She did intensive language courses in 2000 and 2001 in Austria at Deutschintensivkurs, IFK-Deutschkurse, and another at Sprachkurs, Christian-Albrechts-Universität zu Kiel. She graduated from Penn State in 2001, after a self-described, tongue-in-cheek, "seven short years."

Just after graduation, the Islamic terrorist attacks on the World Trade Center and Pentagon took place. A fourth plane crashed into the countryside of her home state, less than one hundred miles from Pittsburgh. Like the rest of the world, she watched in horror as smoke billowed out of the two skyscrapers before they collapsed to the ground in an explosion of debris and dust. She entered the United States Air Force Reserve Officer Training Corps, graduating from the Aerospace Basic Course in 2002.

In June of that year, she reported to Wright-Patterson Air Force Base near Dayton, Ohio, for work in the Air Force Research Laboratory. According to her resume, her work involved research in a "laser laboratory to enhance satellite sensor protection."

Two years later, she earned the role of executive officer of the Materials and Manufacturing Directorate, "managing the training, evaluation and career planning for more than 100 active duty and reserve military personnel."

It was while she was stationed in Ohio that Shipman decided to embrace her ancestral roots and put on the black shoes of an Irish dancer. Her feet worked in precise movements, choreographed with the dancers around her, while her arms stayed statue-still.

She also did graduate work in German in 2004 as a U.S. Air Force officer, again in Salzburg. She enjoyed hiking in the Alps while there. According to what she posted on Adventure Write, she once lost track of time and found herself locked inside after the Miraball Palace Gardens closed. In trying to find a way out, she stumbled onto a drug deal in progress.

"Thirteen years of gymnastics paid off in one spastic, wall-vaulting, gazebo-climbing, iron-gate scaling escape," she wrote.

Finally, she was off to sunny Florida's Cape Canaveral Air Force Station, where she managed launch-base activities for a missile defense satellite system "testing hardware for spaceflights," according to *Texas Monthly* magazine. *Florida Today* newspaper wrote that her primary responsibilities were related to satellites launched from Cape Canaveral on Department of Defense missions.

Her heart was touched when she saw the funerals for service members killed in the wars in Iraq and Afghanistan, along with retired veterans, and she volunteered for Honor Guard duty, performing military honors at dozens of funerals in the Space Coast area.

She was assigned to the staff of the Forty-Fifth Launch Support Squadron at Patrick Air Force Base and Cape Canaveral Air Force Station. She dealt with satellites being launched on Department of Defense missions. She had earned the rank of captain by this point.

She lived in a new townhouse with a few roommates, a bike ride away from the beach and just a short drive to the air station, which neighbors Kennedy Space Center.

Those who lived near her described her as "the little general," someone who was kind and funny, *Texas Monthly* wrote.

In November 2006, at a house party for the upcoming space shuttle mission, she caught the eye of shuttle pilot Billy Oefelein.

Astronaut Lisa Nowak's official NASA portrait. Photo courtesy of NASA.

Right: Lisa Marie Caputo enjoys a Slurpee at Charles W. Woodward High School in Rockville, Maryland.

Below: Astronaut Lisa Nowak, who served as a mission specialist, flight engineer, and medical officer on STS-121, waves at the camera in the suit-up room before her July 4, 2006, launch. Photo courtesy of NASA.

The STS-121 crew gets ready to board the Astrovan for their ride to the launchpad in July 2006. Photo courtesy of NASA.

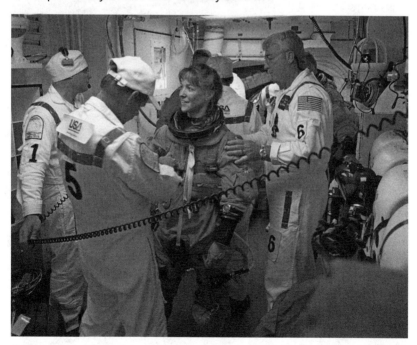

Astronaut Lisa Nowak is readied by technicians for her launch in the White Room, an area high atop the launch platform at Kennedy Space Center. Photo courtesy of NASA.

Right: The July 4, 2006, launch of STS-121—the only Independence Day launch of a manned rocket in NASA's history. Photo courtesy of NASA.

Below: Astronauts Lisa Nowak, Mike Fossum, and Stephanie Wilson enjoy some smiles and camaraderie on board the International Space Station during their July 2006 mission. Photo courtesy of NASA.

Left: Astronaut Lisa Nowak does a somersault on board the International Space Station in July 2006. Photo courtesy of NASA.

Below: Astronaut Lisa Nowak controls the robotic arm during the STS-121 mission in July 2006. Photo courtesy of NASA.

The crew of STS-121. *From left*: Stephanie Wilson, Mike Fossum, Steve Lindsey, Thomas Reiter, Mark Kelly, Piers Sellers, and Lisa Nowak. Photo courtesy of NASA.

Astronauts William A. Oefelein and Lisa M. Nowak pose for a photo before starting special training under austere conditions at CFB Valcartier in Valcartier, Quebec, Canada, Wednesday, January 21, 2004. Photo courtesy of Canadian Space Agency.

President George W. Bush meets with crew members of the space shuttle *Discovery* (STS-121), space shuttle *Atlantis* (STS-115), and International Space Station Expeditions 11, 12, and 13 in the East Room of the White House on Monday, October 23, 2006. Photo courtesy of NASA.

Astronauts Charles O. Hobaugh (*right*), William A. Oefelein, and Kenneth T. Ham, STS-110's CapCom team, monitor prelaunch activity at their consoles in the shuttle flight control room in Houston's Mission Control Center. They were awaiting the launch of space shuttle *Atlantis* at the Kennedy Space Center on the afternoon of April 8, 2002. Photo courtesy of NASA.

Above: Billy Oefelein points to the cameraman as he prepares for his flight in the suit-up room at Kennedy Space Center in December 2006. Photo courtesy of NASA.

Right: STS-116 pilot Billy Oefelein at Kennedy Space Center prior to his December 2006 launch. Photo courtesy of NASA.

Above: Space shuttle pilot Billy Oefelein floats through the International Space Station during the mission of STS-116 in December 2006. Photo courtesy of NASA.

Left: NASA's official portrait of space shuttle pilot William Oefelein. Photo courtesy of NASA.

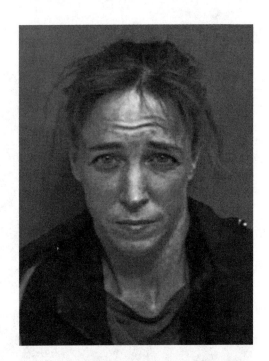

Lisa Nowak's mug shot. Photo courtesy of Orange County Jail.

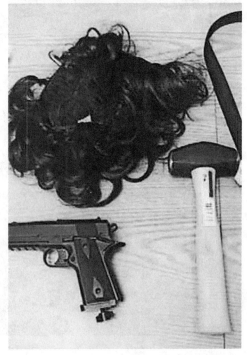

The BB pistol and steel mallet Lisa Nowak had with her when she confronted and pepper-sprayed Colleen Shipman. According to one of her psychiatrists, the black wig she used cost fourteen dollars. Photo courtesy of Orlando Police Department.

Colleen Shipman and William Oefelein walk into a deposition in Melbourne, Florida, on June 23, 2007. Photo courtesy of Malcolm Denemark/Florida Today.

Colleen Shipman is comforted by her attorneys, Kepler Funk (*left*) and Keith Szachacz, after Lisa Nowak pleaded guilty to burglary of a conveyance, a third-degree felony, and a misdemeanor battery charge. Red Huber/The Orlando Sentinel.

15

★　★

Meltdown in New York

Beginning in the mid-1800s, people the world over got a glimpse of the future at one major event every few years. The world's fairs were legendary for unveiling new machines, new technologies, and new ways of living. Their educational, societal, and tourism impacts lasted for decades in the cities in which they were held.

In 1851, London saw an exposition underneath the glittering Crystal Palace. The Eiffel Tower was built for the 1900 event in Paris. And the famous globe sculpture was unveiled at the 1964 world's fair in New York City.

Although the fairs continue to this day, they have evolved into a kind of showcase of nations rather than a peek at things to come. For that, technology companies participate in trade shows all over the world, including *Wired* magazine's NextFest, held each year in New York City.

And if William Shatner is the hottest ticket at a *Star Trek* convention, astronauts are certainly the rock stars of technology events.

Wired's NextFest in September 2006 saw NASA's newly minted space explorers Lisa Nowak and Stephanie Wilson walking like superstars into the Jacob K. Javitts Convention Center in Lower Manhattan. Just two months following their successful space mission, the astronauts were there to talk to techno-nerds about NASA's latest technology—the stuff that wasn't classified—greet people, and autograph photos.

The four-day festival displayed 150 exhibits from companies based in twenty countries. The big buzz at the convention was British businessman and explorer Richard Branson's Virgin Galactic, a commercial outfit designed to take tourists on suborbital flights around the globe.

At NextFest, Branson unveiled the SpaceShip Two's interior design, including La-Z-Boy-style seating for six passengers—all with spectacular views. A 2½-hour tour at sixty-eight miles up will cost each space tourist at least $200,000.

Commercial competition was new to NASA. Although they won the space race against the Russians hands down by landing on the moon, they had yet to take a paying civilian on board a rocket or space shuttle.

Nowak's appearance at the event wasn't unusual. Astronauts are NASA's ambassadors to the world. But her behavior that day can accurately be described as rude, if not downright bizarre.

NASA employee Beth Beck was tasked with looking after the astronauts, making sure they were at the right place at the right time and had everything they needed before, during, and after the event. She had seen to it that several desks were set up that included a stack of photographs of each astronaut for signatures, along with pens.

Nowak and Wilson got themselves settled in, but Nowak seemed bothered.

"Excuse me. There are no pens on my table to sign autographs," Nowak complained.

"Oh, I'm sorry. They're just right inside the desk drawer," Beck told her.

Nowak just sat there, staring at her.

"The pens are in your desk drawer, right there," she said again, thinking Nowak might not have heard her.

Again, Nowak sat staring. Glaring really.

So Beck moved toward the desk and tried to reach over to open the slim drawer, just under the desktop. Nowak wouldn't move. She sat leaning against the desk, making it impossible for Beck to open the drawer.

Beck was stunned. She had dealt with astronauts before, and they were all generally nice people—surprisingly ordinary to talk with in person, despite their high-flying jobs.

Beck walked over to Wilson's desk, pulled a drawer open, grabbed an extra pen, and then put it on Nowak's desk. Then she called her supervisor, Trent Herbert, manager of flight crew operations support at NASA. Beck was audibly upset, calling Nowak's behavior inappropriate.

"Do you want to file a formal complaint?" he asked Beck.

Beck thought for a moment.

"No. I might have to work for her one day," she said.

It wasn't the only phone call he got from support staff regarding Nowak. At a later event, a coordinator approached Nowak. She explained to the astronaut that she had left her purse in a desk drawer next to where Nowak was sitting.

Nowak, again, wouldn't move. She blocked the woman's access to her purse, forcing her to wait until the event was over—and Nowak had left—before she could gather her belongings.

Nowak had displayed "difficult" behavior before. Mike Fossum, Nowak's crewmate on board STS-121, described her to investigators as "prickly" or even "contentious."

"She was not always pleasant to be around," he said. "Our personalities were not a good match."

He recalled being on board the shuttle during their fourteen-day mission. He was struggling to perform a task and asked her for help.

"I didn't train for that," she told him when she refused.

Crewmate Mark Kelly also had issues with Nowak, although Richard Nowak had once told one of Kelly's relatives that he considered Kelly to be his wife's best friend.

"I don't like her," he told investigators. "She was not a good team player. I learned in space that she exhibited bad expedition behavior because she was selfish. She never went out of her way to help any crew members."

But, in late October, Nowak was all smiles at the White House, as the STS-121 crew and their families met with President George W. Bush. Nowak stood to the right of the president of the United States, while Stephanie Wilson stood on his left.

16

★　★

Beginning of the End

In July, following Nowak's flight, she invited Oefelein and his crew to a Houston Astros "Welcome You Back to Earth" event.

In August, he encouraged her to accept a keynote speaking engagement in October with AcademyWomen, a non-profit that supports women alumni, cadets, midshipmen, candidates, and affiliates from the nation's military academies and officer development programs. Nowak asked Oefelein what he thought about it.

"I just walked in! What timing," he said. "I know how you don't like doing these, but this group seems like it would very much benefit from your presence (I can relate!) Anyhow, I would suggest you support if you are up to signing on as the keynote speaker."

That month, she also sent him a picture of her and her crew receiving their Space Flight Medals, along with Hawaiian leis, at a ceremony.

He asked her opinion on booking a hall for his spaceflight reception, a tradition among astronauts to thank the people who have supported them throughout their lives and their training. He preferred the Kurt H. Debus Conference Facility at Kennedy Space Center (KSC), where Nowak also held her reception. It is next to KSC's famed "Rocket Garden," where the space age's earliest rockets stand erect year-round.

"The Apollo/Saturn V center is a little (actually, a lot) out of our price range," Oefelein said, referring to the building that houses a huge Apollo rocket dangling lengthwise from the ceiling and has a ten-thousand-dollar food and beverage fee for use. "Talking to Lisa from the 121 crew and to Joanie [Higginbotham] from my crew, I understand there are several options for food. I don't think we would want a sit-down deal like Lisa, but rather a more informal reception. I will tentatively have 200 or so folks."

In September 2006, NASA records show the couple flew from Houston to Washington, D.C., together for a weeklong trip for Worldspan, a travel technology company that aids in booking airline flights. NASA emails show they were booked on the same outgoing and returning commercial flights for the eight-day trip.

The day after they returned, on September 11, he invited her to a late lunch following a debriefing at Johnson Space Center.

Also in September 2006, Oefelein invited Nowak to train at the gym and on a T-38 jet. Shuttle pilots are required to have forty-five hours of flight training per quarter each year.

"Busy week. Gym days will have to be Tuesday and Thursday AM," he wrote. "I'll need a partner. Plus I need flight time. I still have to get 15 hours. Maybe next Friday evening or next weekend? If you can go, let me know and I will get a jet Friday afternoon."

October saw the annual Astronaut Reunion, held in 2006 at the Westin Oaks Hotel at the Galleria in Houston, with a golf tournament, technical briefings, an Association of Space Explorers function, and annual physicals planned—many astronauts participate in a lifelong health study. And most astronauts attend the reunion.

At some point that fall, Nowak and the STS-121 crew got new cell phones and numbers. The crew members were all trading their new numbers—everyone except Nowak, who refused to give hers out.

Mike Fossum said she was always on the phone, pursuing "private areas for her conversations . . . laughing and having a good time talking on the cell phone."

The week of November 12–17, 2006, about three weeks before his scheduled launch, Billy Oefelein and the STS-116 crew flew in their T-38 jets to Kennedy Space Center to take part in what's known as the Terminal Countdown Demonstration Test. The crew spent three days preparing for their flight.

Oefelein and shuttle commander Mark Polansky boarded the shuttle training aircraft, affectionately known as "the flying brick," to practice for landing. While the shuttle looks like an airplane, it doesn't necessarily operate like one. Its engines are unpowered during reentry and landing, so there is no aborting the touchdown to circle back and try again if something isn't right—they only get one shot at it for each mission, so it has to be done right.

The entire crew then practiced driving the M-113 armored personnel carrier, which would be used in an emergency at the launchpad.

The crew then took the elevator up to the White Room, where they took part in safety briefings about fire. While most of the general public no longer knows who Ed White, Roger Chaffey, and Virgil "Gus" Grissom are, astronauts are very aware of the Apollo 1 crew and their fiery deaths on a Cape Canaveral launchpad on January 27, 1967.

That November, the crew were back at the launchpad the next day to practice getting into and out of the slide-wire baskets that would carry them quickly 1,200 feet to safety on the ground from the White Room 195 feet up. They also inspected the payload bay of *Discovery* one final time.

On the last day, they did everything they would do on launch day, including eating a big meal, putting on their pumpkin suits, riding out to the shuttle, and climbing inside to take their seats. They did a main engine and cutoff exercise while on board.

It was during this week that a house party was held for the space shuttle *Discovery* team and the support crew from Cape Canaveral Air Force Station. That's where Oefelein spotted a thin, pretty, funny, and energetic blonde. Fifteen years of gymnastics and Irish dance had kept Colleen Shipman lithe. His regular workouts at the gym and cycling with Nowak had kept him fit, as well. Oefelein and Shipman's attraction to each other was instant and intense.

"He was just a charming, handsome, very polite man, and he started talking to me," Shipman later told ABC News.

At first, he told her that he was a fire jumper—a firefighter who jumps out of planes to fight forest fires in remote areas. Then he finally admitted what he really did.

They found that they both enjoyed the outdoors. She grew up in Pennsylvania and he in Alaska, so they both loved snow activities, as well.

While Oefelein and Shipman were quickly falling in love, Nowak was on a NASA goodwill trip to England with her STS-121 crewmates. On that trip, crewmate Mark Kelly noticed that Nowak "seemed more withdrawn and kept to herself." That month, Nowak's neighbor in Houston, Bryan Lam, had heard dishes breaking in their perfect home at the end of a cul-de-sac. Lam later told the Associated Press that the police had come, but the Houston sheriff's department has no record of responding to the Nowak home in November 2006.

Oefelein's romance with Nowak was fading. But he was still chatting with her via emails and sending her his schedules.

"I guess these are the last ones" before his space shuttle mission, Oefelein wrote in mid-November as he sent his schedule for the last two weeks in the month.

"Looks like it's working!" Nowak wrote to him on November 27 at 8:00 A.M. She told him she hoped his physical and launch simulation went well. He was also shifting his sleep schedule ahead of the launch to put him into the routine he would be in on board his shuttle flight, when astronauts awake in the middle of the night, Houston time.

Oefelein had suddenly been avoiding Nowak's calls and not regularly responding to her emails. It seemed to her more than just the usual pre-flight busyness.

In early December, days before his scheduled launch date, they exchanged emails about his mission and flight, including his schedules, snafus, and work issues.

"Thank you for the note," Oefelein wrote to Nowak on December 1. "The [simulation] went well. Got a 'V' for victory at the end . . . fueling ran into a problem. Expecting to hear more about it. Off to Med refresher. I hope you can find some nice food and I do hope you enjoy yourself. Let me know how things are going there, OK?"—possibly referring to her situation at home.

She wrote back that "things are going fine" and talked about presentations and events she had to do or attend.

"I am sure you are very busy," she wrote, possibly hinting at not hearing from him much. "I hope the . . . fueling is only a minor glitch and you can look forward to an on-time launch."

On Saturday, December 2, he wrote to her about getting up to date on shots and working out with one of his crewmates, Robert "Beamer" Curbeam, who had also served as Nowak's supervisor when she worked in capsule communications, or CAPCOM.

"It's been nice," Oefelein wrote. "It works to keep the juices flowing until bed time. We had what Roman [Mark Polansky] and Beamer call a 'PC party' here last night. Several folks with PC badges came by and we had a social. Missing one important PC-badged individual, though."

Despite the fact that Shipman was invited to crew quarters, was he hinting to Nowak that he wanted her there? He signed off on the email, "Be safe, Bill."

On Tuesday, December 5, Oefelein emailed Nowak his schedule for the day before launch and launch day, Thursday, December 7. But, unlike Oefelein before her flight, she wasn't invited to crew quarters or the

Beach House to see him off. Instead, the day before his launch, he spent time with Shipman at the Beach House, a special place for astronauts to be with their family and loved ones. It sits on the beach, just past the launchpads.

On December 7, 2006, Oefelein took the same trip out to the launchpad that Nowak had made five months earlier.

"It was dark outside as we approached the shuttle last night," Oefelein wrote in his astronaut blog. "The only thing one could see was this big majestic spacecraft illuminated against a moonless sky. It was an incredible sight. As we stepped out of the van, you could see the shuttle venting. You could hear it creaking and moaning. It was as if it was alive and longing to leave the bounds of the launch pad. I'm certainly not a poet, but I sure felt some serious inspiration last night."

With five minutes left until liftoff, the launch had to be scrubbed because of a low cloud ceiling that brought a cold front with it.

Two nights later, December 9, 2006, space shuttle *Discovery* lit up the night sky just before 9:00 P.M. as it launched from Kennedy Space Center. Shipman was with Oefelein's parents, watching from nearby Cape Canaveral, as her new love piloted the shuttle into space. She could feel the powerful sound waves pounding through her and could barely look at the blinding light of the rocket as it pierced the sky.

"GO, BILLY, GO!!" his dad shouted.

"The launch was 10 times more exciting than I thought!!" Oefelein wrote to Alaska students in his astronaut blog. "The sights and sounds and sensations were, literally, out of this world! You shake a lot when the solid rocket boosters ignite. Then you get pressed into your seat pretty hard as you accelerate up to the speed you need to stay in orbit."

After being in space for a day, he could write back that "Space is fun! I quickly adapted to the zero gravity and have had no problems eating or sleeping. It's fun to eat 'upside down' and sleep on the ceiling."

Like all first-time astronauts, Oefelein spent time in awe of Earth below him and the stars above.

"The sights are incredible," he wrote. "Whenever I can, I try to take a peek out the window. There is always something to see. We've seen thunderstorms, city lights, the Northern Lights, rivers, jungles, deserts, oceans, and so much more. It is quite an experience."

Oefelein and his crewmates were assigned the task of rewiring the entire International Space Station (ISS), adding a new piece of hardware to its truss, conducting an almost unprecedented four spacewalks, and, as with any crew, transferring a lot of gear from the shuttle to the ISS and vice versa.

Oefelein continued to blog with students in his home state of Alaska. One student asked how they take showers in space.

"We can't take showers," he answered. "We take sponge baths with a wash cloth, a towel, soap and water. We just have to make sure our soap doesn't float away."

And there was one special moment for Oefelein, as the ISS slipped past Alaska the night of December 18.

"As we came up the Aleutians, I could start to make out Kodiak, then Homer, Kenai, Soldotna, and Seward. Above them all from the angle I was looking was a big bunch of lights that was Anchorage," Oefelein wrote. "Just to the north of that I could even make out Wasilla. So, there I was, on *Discovery*'s flight deck with the cabin dim, looking out at seven cities in Alaska at the same time. It was one of the neatest sights I've seen so far. There's something special about seeing your home from the air and something I found even better about seeing it from space."

*　*　*

As Oefelein was floating in low Earth orbit, back on the ground Nowak's world was shifting off its axis.

After ten years of struggling in her marriage and considering divorce, and two years of simply existing in the same house, her husband had moved out. She later told a psychiatrist that she "was critical of her husband's engaging in leisure activities and [his] accepting only work projects he liked, when he could have been working more diligently to more rapidly further his career." She also said his lack of interest in keeping their three children on a strict schedule she had devised with their daughters' teacher "was also an area of contention."

And then she found out she was not going to be on an upcoming shuttle flight.

Her STS-121 crewmate Mike Fossum told investigators that Nowak was upset she was not chosen for the job because it was probably her last opportunity for spaceflight. In addition, she felt her assignment for CAPCOM was a step back for her career.

Mark Kelly was the one to deliver the news.

"It was obvious that Nowak was not happy," Kelly later told investigators, noting that STS-121 crewmate Stephanie "Wilson was chosen for the position because she was a team player and well deserving. Nowak was not."

She had spent ten years training, had endured the heartbreaking days after the *Columbia* disaster, when everyone wondered if the shuttle would ever fly again, and then waited two years for the birds to go up again. With only a few space shuttle missions left, she knew her astronaut career had come to an abrupt end.

*   *   *

As Nowak suffered in silence, Shipman and Oefelein kept in touch while he was in space, exchanging emails that showed a playful, flirty, and funny side to the couple's relationship—although surely embarrassing to have the whole world later read what the couple thought would be private missives between them as their relationship evolved.

And Oefelein even made an out-of-this-world phone call, dialing her number from the International Space Station.

"I didn't recognize the phone number on the caller ID so I just let the voicemail get it," Shipman told ABC News in an interview. "It was him, and he was calling me from the space station. I didn't even know that was possible and I didn't answer it!"

Oefelein left a message, telling Shipman that "seeing her was his first priority upon landing," ABC reported.

Their writing in space continued on land in the weeks that followed his return to Earth.

Shipman sent the steamiest email on December 21, while Oefelein was still in space and a day before he returned to Earth.

"Will have to control myself when I see you . . . First urge will be to rip your clothes off, throw you on the ground and love the hell out of you," Shipman wrote.

But Oefelein didn't receive the email until January and responded: "You write such good notes!! You are the best!!! I love you."

She was beside him at a dinner in the quarantined crew quarters after *Discovery* landed back at KSC, just after dusk on December 22.

"The significance of the invitation was obvious," *Discovery* Commander Mark "Roman" Polansky told investigators. "Because each crew member was allowed only one guest."

They parted that night, saw each other the following week for New Year's, and then made plans to see each other again in late January.

"Lots of love coming your way . . . and kisses and a great big giant hug with my legs around you," she wrote before their January reunion.

In one email exchange in January, Oefelein had sent Shipman a picture of himself aboard *Discovery*. He thought he had sent to her a picture holding an Irish charm he later gave to her, floating in space. But he sent the wrong picture, leaving her to stare at it, zoom in on it, desperate to see her gift when it was in space.

"It's like those erotic hidden picture games that they have at the bar . . . only you're fully clothed in the picture . . . and I'm looking for my charm," Shipman wrote. "Ok, not like those hidden picture games at all, but the thought of you without clothes is pretty nice."

"The charm is floating in front of the picture you boob!" he responded. "What did you say your eyesight is?"

When he realized his mistake, he wrote that Shipman must "really have me around your finger that I can't even function without you here, and with you here, I am slightly smarter than a slug."

By January, the couple was saying "I love you" regularly to each other.

"I have to go someplace cold and snowy . . . arg!" Shipman wrote. "I'm taking your picture with me . . . because I like to look at you . . . and I miss you . . . and you're hot . . . and I miss you . . . and to get through this cold and snowy and lonely night. Did I mention I miss you?"

And he was clearly equally smitten.

"I need to see you—I am having Colleen withdrawals. Must see Colleen."

In the office at Johnson Space Center that he shared with other astronauts from his flight, Oefelein placed a picture of Shipman on his desk—there to be seen by anyone who could get into the building.

*   *   *

With Richard out of the house and formal custody arrangements under way, Nowak wanted to move on with the next phase of her life, one that Oefelein had said included him. She had emailed him while he was aboard *Discovery* and then began calling when he landed. Dozens and dozens of calls and texts from a telephone he had given her, along with calls from her own phone. And most went unanswered by him. Billy-O had stopped calling and emailing her.

She began rapidly losing weight, couldn't sleep, and was obsessing over "fixing" their relationship. Her concentration was so skewed that she had been stopped three times by Houston police for erratic driving.

For the first time in twenty-one years, Nowak didn't send out Christmas cards, bake Christmas cookies, or have her annual Christmas Eve dinner.

She told a psychiatrist that she found out via text message on New Year's Day that Oefelein had spent the night with another woman. She said the next day she and Oefelein talked in person, and he told her that he had been seeing another woman and her for the previous six weeks.

"He was upset," Nowak recalled. "He wanted to be open and honest . . . He was confused about where he was heading with his two relationships and still cared very much for [me]."

A psychiatrist wrote that "Nowak, with her inexperience and innate social naivety . . . focused on his being upset and dropped into her problem solving mode, assuming that it was her duty to help him solve his dilemma and once again become a happy man. Cmdr. Oefelein responded to this by continuing his relationships with both her and the other woman."

Oefelein was also sending mixed signals. In early January, Oefelein and Nowak exchanged friendly emails via their NASA email addresses, working out their schedules to bike ride or train in NASA's T-38 jets. There was talk of going to two parties—a happy hour to celebrate the halfway point of a space mission and a "splashdown" party for Oefelein's shuttle crew.

"Going to this?" Oefelein asked Nowak the evening of Wednesday, January 3, about the halfway party the next night.

"Only if u r!" she answered back a minute later, followed by a message about when they could train on their bikes for an upcoming race.

"Is there a target time for this eve yet?" she asked.

"6pm-ish," he answered.

On January 5, he sent her his schedule for that week, and they worked out a time to fly in the T-38 jets, just as they had been doing for several years.

"Hi, thanks for the updated schedule!" she said. "If you want to fly sometime, let me know . . . haven't been to Roswell yet!"

Oefelein responded that his days were pretty booked, "but we can pick an evening and head out to grab a bite to eat." He suggested the following Tuesday.

"Sure! I may move my ground school to February," she said of a required training, "but flying would be super either way. Thank you!"

That day, he also sent her an invitation to a party for the STS-116 crew, scheduled for Thursday, January 11, at Villa Capri, an Italian restaurant and banquet facility overlooking Clear Lake, popular in Houston.

That month, he later told investigators, he and Nowak ate lunch at his apartment and she also met his children for the first time. For Nowak, it marked an important event for someone who thinks they're involved in a romantic relationship.

But Oefelein also told investigators that, for him, the romance was over. He said that sometime around January 7, 2007, he sat Nowak down to have a serious talk.

"I told her that I had met Colleen and I had fallen in love and I was wanting to pursue an exclusive relationship with Colleen," he later told detectives.

"She seemed a little disappointed, but she seemed accepting of that. The relationship with Lisa at that point, um, was not how it had been before, you know . . . we didn't call. She tried to call me. A lot. I—I didn't."

According to his version, he was doing a classic fade-out. Men who want to avoid a confrontation will often simply stop calling and stop returning a woman's calls. But people with Asperger's disorder, like Nowak, couldn't understand these subtle social cues.

It's not clear when Oefelein told Shipman that he had broken things off with Nowak.

"I asked him, 'Are you sure that she's okay with this?'" Shipman later told Orlando Police. "'Because you know how these things go.' And I said, 'Is there gonna be some crazy lady showing up at my door trying to kill me?' And he said, 'No, no, no, she's not like that. She's fine with it. She's happy for me.'"

But he also told investigators that at some point he was walking with Nowak, and Shipman's name came up.

"Mrs. Nowak became visibly upset and left Mr. Oefelein," a detective wrote in a report. "A few minutes passed and Mrs. Nowak returned to Mr. Oefelein and apologized for the incident."

* * *

On January 17, Oefelein and Shipman began planning for Shipman to go with him on an astronaut tour of Europe, including Sweden, Norway, Denmark, Germany, and the Netherlands.

Shipman wrote that she would probably be able to go unless her boss, whom she referred to as a "jackass," wouldn't let her.

On January 20, Oefelein had cashed in some mileage to buy her a flight from Orlando to Houston and back the first weekend in February for her birthday. As they planned, Nowak began snooping . . . or, in legal terms, stalking. She logged onto Oefelein's home computer—according to an assistant state attorney in Orlando, he left his password in his desk—and she found the string of romantic emails between Oefelein and Shipman, printed them out, and left them in her car. She also found the flight itinerary for Shipman.

That same week, Nowak said, Oefelein had picked her up from the airport.

"I mean, you just don't do that for anybody," Nowak later told a detective.

Crewmate Mike Fossum told NASA investigators that Nowak had been scheduled to fly with Ron Garran to Washington, D.C., for astronaut Paul Lockhart's retirement party in January. Lockhart, Fossum said, was a good friend of Nowak's. Instead, at the last minute, she cancelled and "jumped into a jet with Oefelein."

On January 23, Oefelein and Nowak attended a meeting for a bicycle race in which they were planning to compete in April. The meeting was at Saint Arnold's Brewery in Houston between 6:00 and 8:00 P.M. Following the meeting, Oefelein later told detectives the pair went their separate ways. It was the same day she printed out a collection of maps and directions from Houston to Orlando.

A January 29 email from Oefelein to Shipman shows their planning for the April European trip. He signed it, "Love you."

Nowak's psychiatrist would later write that, at this point, "Nowak was acutely depressed and hyper-focused on her perceived need to fix Bill's problem."

She was sleeping, at the most, two hours a night and felt physically and mentally exhausted. "After about 10 days and nights of acute sleep deprivation, she had an odd sense of not needing to sleep and having additional energy, despite her physical and mental exhaustion. Her thought processes were becoming accelerated, but inefficient with increasing difficulty in concentrating and making rational decisions. She was beginning to enter a mixed manic and depressive state."

She told the psychiatrist that Oefelein continued to see her during this period and "became sullen whenever Capt. Nowak questioned him regarding his talking to and being open and honest with Colleen concerning his feelings and being involved with two women simultaneously . . . Nowak mistook his sullenness as unhappiness caused by his not being open and honest with Colleen. Captain Nowak was determined that Colleen know of their relationship and for Bill Oefelein to have an open and honest talk with Colleen so he could decide what he wished to do with his relationship with each woman."

On January 30, Oefelein took an overnight trip to Stennis, Mississippi. He told investigators he might have gone on a morning run before he boarded a commercial flight to Mississippi. He said that either on that morning or in the days before Shipman's February 1–4 visit, he saw Nowak at work.

"I hope you have a nice weekend," she had told him, knowing that Shipman was coming into town.

A January 31 email from Oefelein to Shipman showed he was planning to take his new girlfriend to Alaska on a March trip that involved ConocoPhillips hosting the entire STS-116 crew and their guests.

"They want your size for the arctic gear for the snowmachine outing. I think I can figure that out—sized sexy and athletic," Oefelein said. "ps—Due to noise requirements, I have asked Gina to get us a room at the Captain Cook hotel also. Usually, I stay with my parents on these trips, but we need some 'privacy'!!!!"

17

★ ★

Un, Deux, Trois

"So you brought a toy gun?" Becton asked Nowak.

"I didn't even have that out," Nowak said of the BB pistol.

"It doesn't matter—you brought it with you. Why would you have it in your purse?"

"It, at one point, I just, I thought, if she wouldn't talk to me and if it looked like something that was real, then it would be persuasive. Obviously, I, I, the logic part does take over at some point and none of that stuff, I mean . . ."

"Okay, what if that didn't work?"

"What if what didn't work? The gun?"

"What if the gun didn't persuade her to talk to you—then what?" Becton asked.

"She just, she refused to talk to me about this. Then the only other thing I could have done is gone back and requested to make it a three-person discussion, which I didn't want to do because . . ."

"Because of the outcome?"

"I didn't know the outcome. I just wanted, like you do, like you would like to talk to me without me talking to her. I wanted to get the uninfluenced version of what she thought about me. Without it being now changed because this conversation was probably gonna happen."

"So you're afraid of the influence that he might have on her? Or she might have on him?"

"Either one. I wanted to be able to discuss it without . . ." Nowak said.

"Discussing it with him face to face wasn't enough for you?"

"Discussing it with him?" she said, puzzled.

"Yes."

"But I don't know what he discussed with her."

"Well, what about the discussion you have with him?"

"He had a companion in dating kind of way, that she was free to be that, whereas, obviously whether or not he wants to, I'm at the moment not able to act in the same way," Nowak said.

"That wasn't enough for you?"

"No. We still do some things together."

"On a personal level, at work or both? So he tells you one thing, but his actions show you something else?"

"It's possible."

"Which would leave confusion on your part as to where you really stand?"

Becton felt a little sorry for her—*Oefelein had obviously said they could still see each other—and sleep together. This astronaut had been reduced to nothing more than a booty call for this man, and she didn't even realize it.*

"I can think of the word confused, but I guess that's more along the lines of pointing to if she knows more along the lines of if she had any idea. I need to know where she stands. If she even had any idea that I fit in or not. Or whatever."

"Does it really matter what she thinks? 'Cause he told you where you stand," Becton said.

"He told me where she stands. But that doesn't say where I stand, or where I want to stand."

Oh wait a minute, Becton thought. *The open relationship was* HER *idea—not his?*

"Where do you think you might stand if you found a companion?"

"Given the right conditions, one could have a number of companions now. You take that to a more serious level at one point, most people don't wish to have multiple companions that are very serious."

"Did he use the word companion?" Becton asked.

"He did use the word companion."

"What does companion mean to you if you had to define it?"

"I think in this case, it means to go out with people without wondering if it's okay, without issues about it, doing things together."

"So, you're not really a companion with her?"

"Am I a companion with her?"

"Let me rephrase that—with him—you were the companion, but now you're not?" Becton asked.

"I still feel like I'm a companion, not if I'm not one," she said. "I mean, you know, we go on bike rides and other events. As far as someone to do them with, whether there are other issues, I guess I see that better now."

"That doesn't irritate you, upset you, or cause you concern?"

"It would cause me concern if she was aware of my . . . I can't find the right word . . . people should define what they want to do, how they want to run their lives, and who they want to spend time with. And some cases . . . at first don't criticize it. That means spending time with a number of people, at several different levels. But if you don't have that information to make that decision, decide if you want to do that, then, yes, that's keeping something from somebody."

Okay, so maybe Oefelein had told her he wanted to see her AND Shipman?

"I don't know that it would have been my position—if that wasn't the case—to give it. I guess it was worth trying to find out if it needed to be given, then it may have been more appropriate to then go back and say, 'Well, I think there's a number of things that you guys should discuss, or we should all discuss.' Whatever—but I do know that part was, I guess, probably something that is very important to myself. I think that was very important. That's all."

Becton wondered, *Did she come here to find out anonymously if Shipman was open to dating more than one person at a time? Or did she come here to eliminate the competition?*

"What are you worried about?" he asked.

"It seems like she was aware of my name and who I was, from a working standpoint, and then you said that she put two and two together and then figured I was . . . jealous?"

"I don't know, I didn't ask. I just asked if she knew you," Becton said.

"She told you something about me that was more than astronaut stuff?"

"No, she didn't. You know, the problem is and, I'll spill it all out. The problem is this—you're not understanding the relationship that's going on between you and Bill. And you really question whether or not he's telling you how he truly feels because she's involved. The problem with this is, is the fact that you're reading more into the whole situation than what really exists."

"That I'm reading more into the situation than him?" Nowak asked.

"With both of them."

"But what that means . . . between the two of them—what do you mean?" Nowak asked.

"Between the three of you and you can't process that," Becton told her. "It doesn't process for you. It doesn't register. So, you are of the opinion that she's giving all this information about you and she hasn't. As a matter of fact, the only information I have is from the time you started following her before she got on the bus to the time that you pepper-sprayed her. Now you can choose to believe that, or not. But that's the truth. And the matter of the fact is, you're trying to protect people and hide the whole story, 'cause you don't want people to get hurt. And you know what—you've crossed that line, there's no going back to that. It's done—finito. You need to decide now, what you need to worry about now is getting the help you need. Well, I've been lending a hand out to help you, but you don't want to take it."

"And you were telling me the truth?"

"Yes."

"All of the truth? Everything? What you want from me? But you won't give it to me."

"I can't give you one hundred percent of the truth, 'cause there's some facts I need to keep to myself to make sure you're telling me the truth. I wouldn't be a detective if I did otherwise."

"Facts concerning this relationship?"

Good Lord, she sang the same song over and over, he thought.

"No—facts concerning what's happened."

"Okay. I'm not sure . . . it's not my place to directly do something," Nowak said. "She should be able to make decisions knowing, having all the information available to her, just as I would like to. I'm not sure either of us do."

"By information—if he was going to see—which one of the two of you, he was going to see and be with?" Becton asked.

"Well, obviously, that was. I didn't get the chance to do that. I mean the time. I see the times that I spent. I was there, but I don't know . . . I got . . ."

"I mean, 'cause you never . . . You put a lot on the line into this relationship with him."

"That's why I would like to have gotten the information," Nowak said. "I understand that she's desperate about it—it's because I understand that."

"Okay—so this is coming together. The two of you had a sexual relationship at one point. And you would like it to turn into something else?"

"OK, you're not quite accurate, but go ahead."

"And you're struggling with the fact that you and your husband have split up. And here's a guy that has shown interest in you. But you also have three children to think about. So you can't let just any Tom, Dick, or Harry come into your, your life because when you do that, it's, you not only committed your life, you've committed the life of your children."

"That would be true."

"And you know, if this guy isn't going to—if he's not going to be exclusively you and your children—because you don't want your children seeing men coming and going out of your life, then you won't have anything to do with him . . . romantically," Becton said.

"Yes, that's true."

"The problem comes from the fact that you find out he's been seeing Colleen."

"It wasn't kept from me."

"No. But he wasn't telling, he didn't bring it . . . he didn't tell you. You found out. They weren't keeping a secret, but they didn't, he didn't come out and tell you. He wasn't forthcoming in setting out the entire blueprint of what was going on?"

"I don't know what the whole blueprint is. I didn't, like, find out separately or by accident. He told me that he was going out with her."

"So then why'd you need to know where you stand?"

"I don't know if what—to what degree he—or whatever—how serious his plans were for developing something further. And I also don't know what she thinks about that or about if, if she believes that I have potential. I don't know either."

"But if she's going out with him, why would you have potential?"

"She may not believe I do. I don't know that."

"Goes back to why would she think that you have potential if she's going out with him?"

"Define going out," Nowak said.

"They're dating. If she's sleeping with him, why would you have potential?"

"Well, that's what's going on then. It doesn't necessarily mean that some people do that without intending to make a lifelong commitment of it."

"Would you wait for him?"

"Would I wait? Wait for what?"

"For him to finish with her?"

"I don't know that he was in a position of exclusively being with one person. I don't know that."

"Okay—so if he was going to date both of you, would that be acceptable to you?"

"As long as everybody knows that that's the case. As long as everybody is aware of that," Nowak said.

"As long as everybody knows that he's dating two women, you're okay with it?"

"I mean, a single person can go out with as many people as they want."

"And you're fine with that?"

"Well, it depends," she said. "I think for a time, I think it's possible for a period of time for people to go out and be with other people and not give it a thought. Now, if somebody gets to a point of wanting to make a commitment, whether, you know, maybe there's five people involved. If any one of them wants to be more serious, then, yes, something has to be done about that."

"You did have sex with another man. Your husband was cheating, am I correct? Okay, maybe I'm not.

"Well, it's not pertinent."

"It was pertinent," Becton said. "Let me go back to the fact that you're not in a position to decide what's pertinent or not. If you don't have an answer, just say you're not gonna answer, okay. I'm not gonna ask a question that I don't have point to, okay."

"Okay.

"So did you cheat on your husband?"

"In the context of what?"

"Physically cheating."

"Right."

"And with your answer, I'm gonna say that there have been times that you allowed certain thoughts about other people to go beyond what they should have gone."

"Yes."

"Which shows you're very committed. That you stick to one thing."

She mumbled something.

"Look at your career."

"I try not to think of it," she said, knowing that she hadn't been picked for a second mission.

"You have thoughts of being less than faithful to your husband, but you didn't go through with it. You have a career that you stick with through the good times and the bad and then there probably have been times you thought of just getting out of it, but you didn't. Which means that you're the type of person that believes in having a goal and going for it."

Becton would find out the next day from Oefelein himself that the couple did, in fact, have an affair.

"I believe that people are important," she said. "My work is not the most important thing. I mean, that's why, even though I learned years and years ago that things weren't working out at home, you know, it's like I didn't want to hurt people."

"Do what you can to make it work?"

"Yes, yes."

"That's commitment."

"I'm not sure what you call it," she said. "I mean, you can ask him, as far as being committed to each other—in a hard-life way. I mean, that wasn't there for a long time. But, probably thought people . . ."

"But for the kids, you guys stuck it out?" Becton asked.

"I tried for a long time. Now, in that position, though you're not really free to go beyond a certain point with other people. But you may enjoy other people's company, or spend time with them. And, I'm sure, if I had a chance . . ."

"Well, with that said, how can you tell me you'd be okay with this guy being with both of you at the same time?"

"Well, if that was to be the case and everybody was in a position to do that, and okay with it . . ."

"That's not what I asked."

"No, because we weren't . . . I wasn't quite in a position to be that," she said. "I mean, not yet."

"Why weren't you in a position to do that?"

"'Cause I felt it was inappropriate to—finish what we were in the process of doing something . . ."

"So, you wanted to wait until you were divorced before starting a relationship?"

"Well, doesn't that make sense?"

"So what did you want to find out from Colleen?"

"What she believes their status was."

Nowak told Becton that she and Oefelein had a bike ride scheduled for Saturday morning but had to cancel it because of Shipman's visit.

"Are those intimate times for you?" he asked.

"Bike rides?"

"Yeah."

"Those aren't intimate. Personal?" Nowak said.

"What's the difference between personal and intimate?"

"I guess the same," she said. "Well . . . affectionate . . ."

"Okay, so there's personal between the two of you, at least for you?"

"I was told it was not just me," Nowak said.

"He's told you that they're personal for him, too?"

"He didn't. I think that you have this impression that I'm going in a direction and giving no input at all that, you know, I just wanted something to happen," Nowak said. "And that this is all my idea, and that he has no indications toward me—is that what you were thinking?"

"No, I told you what I was thinking earlier. I was thinking that the two of you were building on something, working on something. Some sparks are flying and she comes in and interrupts that," Becton said. "And he's giving you mixed signals. He's telling you that, well, he's exclusively for her, but what he's showing you is you're, she's not exclusively, there's you, too. And it's sending you mixed signals. That's what I think. What I'm trying to put together is finding out . . ."

"But does she have any ideas about it?" Nowak asked.

"I would have to say no . . . ," he answered.

18

★   ★

Making Plans

It's not clear when Nowak first came up with the idea to confront or even harm Shipman. But she was clearly making plans. On January 8, she went to the office of United Space Alliance publicists who handle astronaut appearances. After a flight, the first few months are busy with public appearances as NASA tries to create goodwill with the public by sending their best and brightest into the spotlight.

Nowak knew publicist Renee Walker, who later described Nowak to investigators as having "a strange personality."

"When Nowak was pregnant with twins five years before the incident, she would frequently discuss pregnancy issues with Walker because she was pregnant at the time, as well," investigators noted in their report. "However, when Walker passed Nowak in the hallways now, Nowak avoided eye contact with her."

Nowak asked Walker for a copy of the calendar that would list all activities, events, and work details of the STS-116 *Discovery* crew. Walker told her she didn't have it yet.

The next day, Walker was walking down a hallway with Dana Davis, a publicist who worked for United Space Alliance, a private contractor for NASA. Nowak approached them and, again, requested a copy of the *Discovery* crew's postflight calendar. She was again told it wasn't ready. Later that same day, Nowak asked Walker a third time for the calendar.

She was like a pit bull with a meaty bone, chewing on the question over and over.

Two days later, on January 11, Nowak requested the calendar from Walker and was told it still wasn't ready, but she had a handwritten version. Nowak said that would be fine and was given a copy of the six pages for January through June. Walker reviewed the notes with her because some of her handwriting was illegible.

"I was concerned about her persistence and discussed the issue with

Renee [Walker]" Dana Davis later told investigators. "During our conversation, Renee told me she 'heard Lisa was having an affair with one of the crew members.'"

"It was unusual for Nowak to be so persistent about the calendar," Walker said, "but sometimes astronauts like to support each other, so perhaps it was not extremely unusual." She never asked Nowak why she wanted the calendar.

Bob Curbeam, Nowak's supervisor in capsule communications (CAPCOM) and Oefelein's crewmate, saw her in early January and noticed she had lost a lot of weight.

"I asked her if everything was 'okay,'" Curbeam would later tell a Florida judge in a letter. "Each time, she said that things were fine, but I noticed that she simply did not seem to be her usual self. The fact that she has always been physically fit and [an] active individual, yet was experiencing sudden weight loss and displayed out-of-character preoccupation with something that she was unwilling to talk about made me feel as though all was not well."

Her STS-121 crewmate Piers Sellers, who shared an office with her, also noticed that she was withdrawn and had lost a lot of weight.

"Also during this period, she began to look extremely fatigued and stressed," Sellers would later write to a judge. "I asked her a few times if anything was the matter, and each time she indicated that she would rather not talk about herself further."

On January 12, Nowak logged in to her work computer in the astronaut office building at Johnson Space Center and went to the MapQuest site to get directions from Houston to Orlando International Airport. She printed one set of directions, using the Florida Turnpike to reach the airport. She wrote on a sticky note attached to the airport webpage printout that she would "need change and small US for tolls."

But then she changed her mind. She wrote on the original map to "redo no toll." She couldn't take a chance of being seen or having her license plate recorded on a camera at a toll booth.

She also printed out a map of Cape Canaveral, Florida, and the area where Shipman lived, which was so new it wasn't on GPS maps.

On January 23, 25, and 30, she printed more maps, enlarging ones of Sumter and Lake Counties, north of Orlando, to find a back way to the airport using county roads. The directions had her exiting Interstate 75 at Wildwood and taking State Road 44 to US 27 South, then onto State Road 50, which runs into the heart of Orlando.

She later told a psychiatrist—which was validated by some of her astronaut colleagues—that, despite being a Naval navigator and an astronaut, she has a terrible sense of direction and that is why she printed out so many maps.

She decided to drive to the La Quinta Inn, just off airport property, and take their shuttle to the airport. That way, there would be no record of her parking at the airport. She jotted a note at the bottom of a page she had printed out with the hotel's address and phone number. Her note read "Airport Shuttle 5am–11pm."

On January 28, six days before leaving on her fateful trip, she had gone to her local Sports Authority and bought a knife and a BB pistol, along with ammunition and a can of pepper spray, paying nearly one hundred dollars in cash for all of it. Receipts found in her car showed she went through the checkout line a second time to buy sports bras and shorts, paying for those items with her credit card.

That last week in January, at Oefelein's apartment on his USS *Nimitz* aircraft carrier stationery, she had hastily scribbled Shipman's flight itinerary, found in an email when she went through Oefelein's computer. She also noted Shipman's unlisted home phone number on his phone bill and the hour Shipman would arrive back in Orlando on a United Airlines jet late Sunday night. She printed out the steamy emails between Oefelein and Shipman, later found in her car.

On January 29, Nowak submitted a request for scuba training with Oefelein to Erlinda Lee Stevenson at Johnson Space Center (JSC). She also requested an all-day T-38 flight with Oefelein on Thursday, February 8.

"Ultimately, I had to flip-flop crew members on a simulation to get Oefelein and Nowak free for an all-day flight," Stevenson told investigators.

The T-38 flight on February 8 conflicted with an all-day simulation Oefelein had at JSC. Stevenson took Nowak off a half-day simulation on Wednesday morning and scheduled her for a full-day flight with Oefelein on Wednesday, February 7. Stevenson did not know the destination of the flight.

"Both needed the flight time and it was not unusual to juggle assignments and training . . . However, all recent training requests with Oefelein and Nowak were initiated by Nowak," Stevenson said.

A prosecutor and Detective Becton would later speculate that Nowak wanted to kill Shipman, dump garbage bags filled with her dismembered body in a swamp, and then use the training flight to console Oefelein when he hadn't heard from his girlfriend for a few days.

Shipman arrived for that long weekend visit on Thursday, February 1. Oefelein picked her up at the airport.

On Friday, February 2, astronaut Mike Fossum, who had flown with Nowak on the July shuttle mission, saw her in the office in the afternoon. "She was bustling around with papers. She seemed happy, cheerful, and smiling, but she never made any mention of a long-distance road trip," Fossum said.

On Friday, Nowak worked until 7:00 P.M.

On Friday, Oefelein and Shipman spent the day in Galveston, checking out a helicopter training program at the airport and then eating lunch overlooking the Gulf of Mexico. They went to the movies that night and saw *The Messengers,* a horror film about a dark force that invades a North Dakota sunflower farm.

At home on Saturday morning, February 3, Nowak packed her black duffel bag. Like a conscientious NASA employee with strict attention to detail, she had made a list of what she needed for this trip: knife (she wrote "sharp" beside it), BB pistol, ammunition, hammer, plastic gloves, and a disguise, including a black wig, trench coat with hood, and glasses. And a lot of cash to pay for gas and hotels. She wrote it all down on her astronaut "flight events/history/briefing" stationery.

On Saturday, in the bedroom of Oefelein's Pirate's Landing apartment #1115, Shipman saw two bikes, including a purple one.

"Since when do you ride a purple bike?"

"Oh, that belongs to someone on the bike team with me."

Oefelein and Nowak were planning to compete in the MS150 Bike Ride in April, so they trained together and worked out in the astronaut gym together. He told investigators he considered Nowak "one of my best friends at NASA."

"Well, whose bike is it?" Shipman asked.

"It's Lisa's."

Shipman thought about it.

"Do you think it's wise to keep her bike here if you two have split ways? I mean, I'm not going to tell you to get rid of it, but it makes me uncomfortable. It makes me want to pull away from you. It makes me think you're not quite ready to cut ties with her."

"I'll get rid of it tomorrow," he said. "That bike is gone. I have made it clear to her that we are just friends, and she just keeps it here because we are on the same team."

That night, Shipman and Oefelein went out to a club to celebrate the birthday of one of his colleagues. After a couple of drinks, they returned to his apartment and were lying in bed. And that's when Billy Oefelein called Colleen Shipman "Lisa."

That same night, in DeFuniak Springs, Florida, Lisa Nowak was lying in bed alone in room #118 of a Days Inn, thinking about what she was going to do.

"Her deranged thought processes focused only on how she would get to Orlando without Bill or others she knew spotting her," her psychiatrist George Leventon later wrote. "Going by military or NASA air transport was out of the question. She reasoned that commercial flights were too risky since she often flew and met people who knew her at the Houston airport . . . Capt. Nowak decided to drive and pay for all of her expenses in cash in case Bill later looked at her credit card records and realized she had gone to speak with Colleen and misinterpret this as her being jealous of Colleen and his relationship."

And, he wrote, Nowak had no plan as to what she would do once at the airport, "other than disguising herself with a cheap $14 black wig."

19

★ ★

The Attack

It was after 1:00 A.M. on Monday, February 5, and Colleen Shipman was tired. It had been a long flight from Houston to Orlando, and she was making her way to baggage claim. She hadn't noticed the woman in a dark jacket, with dark hair and a hat, walking behind her.

Adding to her misery, her bag was lost.

"It's coming in on the next flight, Miss Shipman," the attendant assured her after she waited for about half an hour.

"What time will that be?" she asked.

"Around 3:00 A.M."

She was irritated, but there was nothing she could do. They gave her a twelve-dollar voucher for food, so she went upstairs, bought hot chocolate, water, juice, and a pastry at Starbucks, and sat in a back corner to eat. Then she wandered around the Orlando airport, peering into closed store windows. The petite woman looked inside the Kennedy Space Center Store and spotted the flight suits like the one her boyfriend wore; that made her smile. He looked so sexy in it. And out of it. She tried to call Billy, but her cell phone had died.

She wandered down to a bench beside the moving walkway. She was tired, so she lay down, tucking her carry-on bag under her head for a pillow, and fell asleep for an hour.

She still hadn't taken any note of the woman—who had changed from a jacket into a trench coat—following her. As Shipman napped, the woman walked over to Starbucks and, at 2:07 A.M., ordered a tall light mocha frappuccino.

An hour later, Shipman got up from her nap, walked past a man vacuuming the vast field of carpet in the concourse, and went back downstairs to baggage claim. The woman following her was trying to put on a scarf while walking. She knocked off her hat and left it on the floor as she kept moving.

Shipman found her suitcase waiting for her, so she grabbed it and headed out the sliding glass doors to catch the shuttle to the Blue Satellite Parking Lot.

A blast of cool air hit her when the doors slid open. She looked down the sidewalk to the right and spied the parking spot for the shuttle bus as she inhaled the stale diesel fumes that were trapped on the lower level of airport parking.

The bus pulled up after a brief wait, and the driver got off.

"Let me get that," he said, taking her bag.

"Thanks."

She climbed aboard and plopped into a seat for the ten-minute ride.

The woman who had been lingering in the terminal, the one with the trench coat, ran down the sidewalk, got on, and sat toward the back, carrying a duffel bag. She had the hood of her tan coat pulled down nearly to her eyes. She wore jeans cuffed at the bottom. Shipman understood that. At just over five feet tall, she knew it was hard to find short jeans. In a 3:00 A.M. daze, she stared at the woman's black shoes. Are they sneakers or dress shoes, she found herself wondering? Suede or leather?

But what Shipman found odd was that this woman had on several layers of clothes, like people in northern states do in the frigid days of winter. It's Florida, she thought, and, even though it's February, nobody wears layers like that here.

She looked at the woman's perfect little nose. *I have such a big nose,* Shipman thought. But then she fixated on the woman's glasses—big, round-rimmed glasses, a style popular in the late 1980s.

Somebody's gotta help this girl with her fashion sense, Shipman thought.

The bus wound through the terminal area and then back to the remote satellite parking lots. He pulled into the Blue Lot and asked her in what section she was parked.

"C," she said.

"C," he said back.

"C," she said again and started to giggle. It sounded like an *I Love Lucy* sketch.

At about 3:30 A.M., he pulled up to the covered waiting area, then stopped and helped Shipman with her bag as she left the bus. As he climbed back into the shuttle he noticed that her cell phone was sitting on her seat. He grabbed it and handed it to her. The other woman got off the bus without a word to either one of them.

A light, sprinkling rain was falling, making it feel colder. Shipman was walking toward her silver Saturn in row 31. Leafless turkey oaks were casting skeletal shadows underneath amber-hued streetlights across the parking lot.

Shipman could see the woman in the trench coat, meandering through the rows of cars nearby. *What is she doing? Is she headed to her car?* Shipman began walking faster. She got her keys out. She could hear the "swish, swish, swish" the woman's pants were making as she walked.

Shipman unlocked her car, threw her bag and backpack onto the backseat, and opened the driver's door. She heard running footsteps rushing toward her, frightening her, so she quickly jumped behind the wheel, shut her door and locked the car.

It was the woman in the trench coat. She began slapping at Shipman's car window and tried to open the door. What was she doing? Who was she? What did she want?

"Lady, you scared the shit out of me!"

"Can you help me? My boyfriend was supposed to pick me up and he's not here," the woman said. "I've been traveling and it's late. Can you give me a ride to the parking office?"

Something didn't seem right to Shipman. *Why was her boyfriend picking her up in the parking lot and not the terminal? If this woman needed help, why did she run toward Shipman instead of calling to her? Why did she slap at her window instead of gently knocking? Why did she try to get in instead of talking to her?*

"No," Shipman said. "If you need help, I'll send someone to help you."

The woman in the trench coat wouldn't give up that easily. "Can I use your phone to call him?"

"No, the battery is dead," Shipman said. She picked the phone up and showed it to her through the window. "See?"

Shipman started the car. *This woman looks crazed, like she's on some kind of drug. Maybe she is trying to steal my car or my suitcase or my backpack? Why was this woman targeting me at 3:30 in the morning?*

The parking lot was dark except for a few scattered streetlights, and Shipman was starting to feel very much alone. The woman in the trench coat began crying.

"I can't hear you . . . please help me. Please," she begged, whimpering.

Her instinct was telling Shipman to leave, but she paused. How would she feel if Billy weren't there to pick her up? Maybe this woman really did need help. Shipman tied to crack her window, but the automatic roll-down pushed it farther down than she wanted and she quickly moved to pull it back up.

"I'll send someone to help you, okay? Wait right here."

"Please help me."

Shipman started to say it would be all right, but the woman in the trench coat had pulled out a slim, black spray can and began squirting liquid into the car. It burned. Shipman turned her head and raised her arm to block the noxious spray, then held her breath and rolled the window up.

"You bitch!" she yelled at the woman.

She first threw the car into reverse but then saw there was no one parked in front of her. She slammed the car into drive and sped away from the spot. Her eyes began burning, and her nose began to run. There was a burning in her throat. She saw the exit lane but didn't see an attendant who might be able to help her.

She exhaled, rolled down her passenger and back windows, took a deep breath of fresh air, then pulled up next to an idling shuttle bus with the driver inside.

"I was just attacked by a lady in the parking lot!" she told the driver. "How do I get out of here?"

The driver told her to go back to the E-pass lane and find the attendant.

Marie Smith was working in the toll booth at the Blue Lot when a crying woman pulled up and got her attention.

"Someone attacked me and sprayed something into my car," said Shipman.

"Where!?" Smith asked, looking shocked.

"I was in row 31."

Shipman handed over her parking stub and asked Smith to please call the police. She began to cry softly, as she realized what might have happened if she had let the woman into her car. Her eyes stung, and she began rubbing them. Smith gave her some wet paper towels to wipe her eyes.

Orlando Police officer Timothy Ryan, assigned to the airport, got a call on his radio from dispatch. A woman in the Blue Lot had been attacked by another woman. A few minutes later, arriving at the Blue Lot, he found a thin blonde woman, her eyes and nose red. He put her in his car.

"What did she look like?"

"She was wearing a trench coat with a hood. She had dark hair and glasses."

"What color was she?"

"She was white."

Ryan alerted other officers in the Blue Lot to be on the lookout for a white woman with dark hair wearing a trench coat. He got some personal information from Shipman—where she lived (Cape Canaveral), where she worked (Cape Canaveral Air Force Station and Patrick Air Force Base), how old she was (thirty).

Officer Wendell Reeve was driving through the Blue Lot when he saw a woman who fit the description of the suspect as she walked to the "C" bus stop. She saw him in his marked patrol car, changed direction, and put a black object and a white plastic bag into a trash can then walk away. Reeve stopped the woman and asked her to provide some identification.

Officer Ryan drove Shipman to two suspects they had found in the parking lot. One woman on a bus had on a brown trench coat, but Shipman shook her head no.

Then Ryan drove her to the bus stop where both she and the woman had gotten off. He showed her a woman in a black coat, with her dull brown hair pulled back, wearing blue jeans rolled up at the bottom and black shoes.

"That's her," Shipman said. "She's dressed differently, but that's her."

Reeve searched through the black duffel bag she was carrying and found a tan trench coat with a hood attached. Inside a nearby garbage can, Reeve pulled out a black wig and the white plastic bag—inside the bag, he found a BB pistol, loaded with BBs and set to fire.

The woman, sitting on a bench, tried to reach for her duffel bag, but Ryan ordered that she be handcuffed—for her safety and for the officers.' Inside the bag, they found a steel mallet, trash bags, rubber tubing, and a four-inch hunting knife.

They also found the NASA identification of astronaut Lisa Marie Nowak.

20

★ ★

Murder Kit

"Why did you bring the pellet gun?" Becton asked her.

"I didn't have any of that stuff out." Nowak said.

"Why did you bring the pellet gun? It's a very simple question."

"I was afraid she would refuse to talk to me and I just wanted . . . I didn't do any of those things. I had no intention."

"No. You didn't point the gun at her, but you did bring the gun, and you were afraid she wasn't going to talk to you. So, if she didn't talk to you, you were gonna kill her?"

"No—the thing wasn't even set up to do anything with. I—it wasn't— I just wanted her to talk to me and I was afraid she wouldn't . . . I didn't know if she believed that they met each other and thought that was . . ." her voice trailed off.

"So why the disguise?"

"I believe that she knew who I was."

"Then you think that disguising yourself up, so she didn't recognize you at all is going to help you get into her car and talk to her?" Becton asked.

"That's all I wanted to do—is get in there and then I wanted to tell her what I wanted to talk to her about."

"Why did you write down the ammo nine on the list?"

"I didn't write ammo nine. I . . ."

"So, what was that supposed to mean?"

"I don't know."

"Why would you mean to bring three—only three large garbage bags?"

I believe Nowak had wanted to kill Colleen, use the surgical tubing to stanch the flow of blood, cut her up into pieces, put them in the garbage bags, and then dump the bags into some canal or swamp—there were millions of acres of swampland still left in Florida—and plenty between the airport and Houston.

"To have something to put the trash in, in the car."

"Three large bags to put trash in the car?"

"It's a lot of trash."

"Three? Are you living out of your car?"

"No," Nowak said.

"Are you dumping feces in the bags or something?"

"No. I take trash bags when I travel in the car. That's not illogical."

"Trash bags—no. That's not illogical because [people] have kitchen bags. Kitchen trash-can-sized bags. Lawn and garden bags—that's illogical. Do you really go through enough trash, or go through enough garbage that you need three lawn and garden bags?"

"They just hold liquid better than grocery bags."

"How much liquid are you gonna put in those bags?" *Enough blood to fill a human body?* he thought.

"Well, you have a drink, like chocolate milk, and you don't finish 'cause it gets warm and you put it in there and you don't want it to leak," she said.

"Excuse me for a minute," Becton said—his sarcastic side rearing up. "I thought I had stupid on my head or something. I mean, really?"

"That's all it's for," she insisted.

"Okay, let's put you back in the jury. You're in the jury box and you hear this person brings three lawn and garden trash bags, with ties, to throw away liquid better."

"Is it that important?"

"It is important, okay, because appearance is everything right now," Becton said.

"I don't understand."

"You've got a pellet gun, you've got this list. You've been thinking about this since the end of January—January 23 to be exact. You drive all the way here from Houston to Orlando. You—it's not that you're gonna meet her at her house. You come to the airport. What does that look like to you? You're just coming to talk to her?"

"Yes."

Unbelievable, he thought. "Hay, coño. The damage here. Here's the thing, you've already affected the people you're trying to protect."

"If, if um . . . she decided . . . to say . . ."

"It goes back to it's really not up to her right now—it's the state attorney's office. You have to convince the state attorney's office not to press charges."

"Okay, well first, first thing is probably talk to her," Nowak was begging.

"In a case like this, it carries very little weight."

"Why?" she asked.

"Because it, um, deals with emotions. It doesn't, it doesn't deal with, um . . ." Becton was trying not to tip his hand and say he wanted to charge her with attempted murder. "It's not as much logic as emotions. Okay, the two of you have a similar interest in a, in another person, and because of that—"

"I don't know," she interrupted.

"—interest, you're here."

"Right. I don't know what, if, if, if it's a similar interest—to what degree, or anything. I don't know that—" Nowak said.

Now it was his turn to interrupt.

"It doesn't matter what the degree is, it's just the fact that you have a similar interest. You're here, and because you're here, you're here because you're trying to find out where you stand with an individual— that's considered emotions," Becton said. "Because logic would be for— it's not logical for you to drive all the way from Houston to Orlando just to have a chat with somebody in their car, in the parking lot at the airport—okay. So because it's involving so much emotion, state attorney's office realizes that, and that's why they get the last say in the prosecution . . . They don't want the victim making these emotionally based decisions, after the incident."

"That might be logically . . ."

"To be honest with you, there's no logic in anything that happened tonight . . . What did you need with this?" Becton asked, lifting up the steel mallet out of the duffel bag.

"I don't. Nothing that is . . ."

"It's brand-new. But you were holding this bag when you went up to the car."

" . . . it's sitting in there . . ."

"But you weren't holding it," he said, "so it was very accessible. I mean, why would you need this hammer?"

"I don't need any of it. I told you."

"What would this be for?" he asked, holding up the surgical tubing.

"Exercise band."

"Lisa . . ." He started to say something incredulous, but she interrupted him.

"That is . . ."

"You have an exercise band with a hammer, with the black gloves, with this knife, with this garbage bag and all the cash that you had in here. I mean, really. What's with all this stuff? I mean, this is an expensive knife, brand-spanking-new."

"It wasn't very much," she said. "I didn't have any of that out."

"It doesn't matter if you had it out or not. You had it in your hand."

"I didn't have it in my hand—I didn't."

"So where did you put it?"

"All that stuff was just sitting in there."

"Okay, this bag was in your hand?"

"Yes."

(Her attorney would later try to argue that she didn't have the bag with her at all.)

"Okay, so you had access to this, as a matter of fact, at any time, you could use this hammer to break out the window to get in the car, couldn't you?"

"Well, I guess if I had thought of that . . ."

I think I'll broach the subject now . . .

"Or maybe hit her over the head and try to kill her?" he asked.

"No, no, no, no."

"I mean, this would be great for strangulating somebody," Becton said, holding up the rubber tubing. "Or pick up the parts and suffocating somebody . . . you know? You see where I'm going with this, 'cause this doesn't look good. There was more to your plan than what you told me. You got the makeup on here. You have—it's brand-new, it's not like you're using it."

When she was taken into custody, she had on dark eyeliner that she hadn't used in any of her NASA public appearances.

"So what's the deal?" he asked again. "It's not like you have clothes in there and you're traveling. You have that hammer, and knife and rubber hosing."

"I thought it can be used in this, I didn't, I was trying to think of a way so that she would let me get in her car," she said, admitting to what's known as armed burglary of a conveyance. Burglary doesn't always mean robbing someone. It also means entering into a place—in this case, a car—where you're not invited, welcome, or wanted.

"And I thought if I said the gas was siphoned out of my car that she would give me a ride to a gas station, but it sounded stupid, it was all stupid."

"Okay, so you decided to nix that idea about siphoning the gas."

"I didn't . . ."

"What's with the hammer?"

"The other stuff, I had it all, all this whole thing of finding, what, what could be this whole, get her to talk to me thing, and you know," Nowak stammered. "Like okay, what looks scary and I, and as you could see me, I [acquired] any number of things without any plans, or using any of them. And I have decided . . . I didn't know, I didn't have a plan. I just wanted to be, I really, really wanted her to talk to me . . ."

"Are you gonna tell me where the truck or the car is?"

"Yes, but would you talk to me now? You don't believe me anymore."

"No, not really. Buying a hammer like that, a knife like that," Becton said.

"I had no desire or intentions to actually . . ."

"To what?"

"Cause her to do anything that would . . ."

"So, basically, your intent was just to scare the living shit out of her, basically?"

"At some point, when I was getting those things . . . I mean, I just . . . really . . . I didn't know what I wanted to do," she said.

Then something occurred to her.

"Yes I do, I know what I wanted to do," Nowak said. "I wanted her to talk to me. And I wanted if she, if I, I got in the car and she said, 'No, I'm not talking to you,' then I wanted to persuade her to talk to me."

"Why don't you just say you wanted to scare her to talk to you, 'cause that's what the truth is?" Becton asked.

"Yes, at some point . . . it seems like if it would make sense, but obviously, that didn't make sense."

"This whole thing doesn't make sense," Becton said. "I mean, really, none of it makes sense, does it?"

"No, of course not. This is . . ."

"Insane," he said.

"Is it?" she asked.

"Yes it is—very."

21

★ ★

The Car Search

It was about noon—more than eight hours since the attack in the parking lot and more than five hours after he started questioning her.

"OK, are you going to take me to your car?"

"Yes."

Finally, he thought. Becton felt as if he had been questioning her for days, but at the same time, he wasn't sure where the time had gone. His sergeant was searching airport parking lots for a car from Texas.

I never was very good at chess.

Becton arranged for a marked squad car to take her to her car; he would follow in his vehicle. She still hadn't told him where it was—only that she would take him and another officer to it. He offered to get her something to eat, and she asked if she could have a granola bar out of her duffel bag. He handed it to her and she unwrapped it and began nibbling on it.

Becton followed the squad car away from his airport office, and began hoping Nowak would not rescind her consent to search the car. He wondered what he would use as probable cause—the legal reason—for a search warrant on which a judge would sign off.

I know she drove here, so there should at least be a map in her car. It would be nice if there were a notebook with notes, or pictures of Shipman. I need a lot more than just this attack in order to charge Nowak with attempted murder.

But he knew at this point that he didn't have probable cause for an attempted murder charge. Attempted robbery, maybe. Burglary to a vehicle with a battery? Absolutely.

Becton realized they were leaving the confines of the airport, driving on Jeff Fuqua Boulevard, across the bridge over the Beachline, toward hotel row on Semoran Boulevard. This is the area where all the hotels outside the airport are located.

Nowak parked in a hotel parking lot.

He wondered how she would know the hotel had a shuttle service to the airport—she would need to look up this information on the internet, and that would involve preplanning. . . . indicating premeditation. The pieces of the story were coming into focus.

They turned into the La Quinta Inn parking lot.

Did she get a room here and wait for Shipman?

On the far side of the hotel, facing a heavily wooded area, they pulled up to her husband's BMW, with the faded paint job.

That's not what I was expecting from an astronaut.

He took a breath and prepared for the moment of truth—consent or search warrant?

"Do I have your consent to search your car?" he asked.

"Yes."

Thank you, God!

He rolled down her window in the squad car and told her she could tell him at any time if she changed her mind.

He took a moment to think about the best way to search the interior of the car. He knew he needed to be methodical, keeping track of what he found and where he found it. He started on the driver's side, the driver's seat of the car, but there was nothing on or under her seat. Then he saw a stack of papers on the passenger seat.

Not yet. I am going to stick to the plan.

He went to the rear seat on the driver's side. He found a baby seat on the rear passenger side of the car and, in a garbage bag on the floorboard behind the driver seat, two used diapers. He had changed enough diapers as a father to know what a used diaper looks like. A wave of cold fear washed over him.

Did Lisa Nowak bring a baby with her? Has this baby been by itself all this time? I have to find this baby before I do anything else.

He looked at Nowak and said in his most serious and stressed voice, "Where is the baby?"

Nowak was very calm. "There is no baby."

Now he was very confused as to why there were two used diapers in Nowak's car, a car seat, and no baby. Now he wanted to know why there were diapers in her car—not adult diapers, either. They are Pampers for toddlers.

"I used the diapers to pee, keeping me from having to make any extra stops on my way to Orlando," she told the detective.

"In my entire career, I cannot think of anything else as crazy as urinating in a diaper while driving," Becton said. "So, let me get this straight. A bona fide captain in the Navy, who can fly numerous different combat aircraft, who is a member of the Astronaut Corps, and who has flown in space, around the world, is driving from Houston, Texas, to Orlando, Florida, to see the girlfriend of her ex-boyfriend and is using diapers on this trip?"

What have I gotten myself into? What did I do to deserve this? If I were to bring soiled diapers to my property-and-evidence people, I would get the nastiest note, telling me I had to throw away the used diapers.

In the backseat, he found BBs for the BB gun, the box in which the BB gun was sold, the plastic package that had housed the hunting knife, more empty garbage bags, more surgical tubing, and the Sports Authority shopping bag that Nowak had used to carry the BB gun, the hunting knife, and BBs.

He moved to the trunk and found some NASA documents, and new diapers—there must have been about fifty clean diapers lined up in a row.

This is just really weird.

He went to the front passenger seat of the car and found the gold mine of evidence he had been hoping for: all the maps Nowak had printed from MapQuest, taking her from Houston to Orlando. Along the border of the maps Nowak wrote the mileage and the time traveled. There was a hand-drawn map of Shipman's neighborhood, and instructions for driving from Orlando International Airport to Shipman's house. There was even a map of Kennedy Space Center and Cape Canaveral Air Force Station. Included with the maps was a hotel receipt from the Days Inn in DeFuniak Springs where Nowak had stopped and spent Saturday night. She had used a fake name and fake address on the receipt. She had paid cash for the room. Then he found a letter to Billy Oefelein's mom from Nowak:

Dear Mrs. Oefelein,

Thank you so much for your wonderful, encouraging notes to me! I enjoyed sharing the amazing experience of my mission with you from space in July. The past few years have been both wonderful and challenging in many ways. Bill is absolutely the best person I've ever known and I love him more than I knew possible. Your kindness of supporting us even under such circumstances as have existed in the past is nothing short of extraordinary. Fortunately, that past situation is finally coming to a close with formal separation and separate living arrangements accomplished, and I am in the process of completing all the official divorce paperwork. It is long overdue, but it is finally here and I am very much looking forward to getting to know you even better. It has been my privilege and honor to receive such special caring from you. Especially since my parents are not as supportive right now, it *really* means a lot to me to have another Mom to turn to. Thank you so much for being there for me!

Love,

Lisa

Becton collected all the evidence from the car he planned to take back with him, leaving the diapers behind. He drove Nowak back to the station, on the way thinking about on what charges he had probable cause to arrest her and what state statutes he could prove beyond a reasonable doubt. He knew there was probable cause to charge Nowak with attempted murder, aggravated stalking, and a list of other things. But he didn't see how attempted murder could be proved beyond a reasonable doubt. He and his sergeant called the Orlando Police Department's legal adviser, who told them the best charge was attempted kidnapping and attempted burglary with a battery.

22

★   ★

Behind Bars

After several hours of questioning and the trip to search her car, Becton put Nowak in a holding cell inside the Orlando Airport Police Station. The stark, white room is nine feet by five feet, with a bench not quite long enough to stretch out on. It was about 12:30 P.M. on Monday afternoon—almost twelve hours after Colleen Shipman had gotten off her flight from Houston.

Nowak knocked on the cell door, and Officer Rich Albe opened it.

"Yes, um, I requested a lawyer, please."

"That you [did], but at the moment, we can't do anything about that, okay," Albe said. "You'll have a right to that—all right—and all evidence once you get down to Orange County booking."

"I was told when I first started talking that I had a right to a lawyer," Nowak said, seemingly coming to realize how serious things had gotten.

"Yes, you do."

"Okay, I already talked to them about that, right?"

"You spoke to the attorney? Is that what you're saying?"

"No, I spoke . . ."

"You do have that right, but right now you're being detained, okay," Albe said, trying to explain police procedure to someone who had never been in a police holding cell before. "And you're going to Orange County Jail and that will come about at some other point. We're not going to be discussing it. You'll have an opportunity."

"So by talking before one came, I now . . . " she said.

"I can't get into whatever it is you spoke to the detective about."

"I heard him—he said I have the right," she said, sounding determined.

"Yeah, and you do, but not at this moment," he said. "We're going to go ahead and make a phone call to the jail."

He shut the door.

She moved to a corner of the cell for a few minutes, then stood by the door. She looked like a punished child, standing to face the door, knocking again.

"Do you know how much longer it's going to be?"

"Be patient," he said.

"I told him about a phone call I need to make," she said. "There was an offer to make the phone call for me . . . [about] taking care of my kids."

She had an appointment that afternoon—it was her son's fifteenth birthday, and she was supposed to pick up his birthday cake.

"He [Becton] can make the phone call for you if you give me the number and the person we need to speak to," Albe said.

"Is there any estimate of when I'd be able to make the call? I'd prefer to make it on my own. If it turns out to be a long, long time. I'd like to take this instead," she said of their offer to make the phone call for her immediately.

After Officer Albe left, she stood there, playing with her long hair, picking at and biting her fingernails. She appeared to be softly crying. Then Albe returned.

"Don't quote me on it, but it appears he's got about an hour's work," Albe told her, closing the door.

After a few minutes, Nowak looked like a caged animal and began knocking on the door. Then she knocked again. And she kept knocking until Albe came back.

"Could he talk to me?"

"He's not going to be talking to you now."

"How 'bout the phone call?" she asked, hoping to talk with her ex-husband about picking up the children after school—and picking up their son's birthday cake. She was clearly desperate that Richard Nowak not know where she was or what she was doing.

"Did you want us to make it for you? Or if you don't want to do it now, we can wait," he asked. "You can just knock later and ask. It's up to you."

Albe explained that Becton had at least another half hour of work left to do before he could talk with her again.

"Another question I have is, when we were in the car, we discussed a phone call."

"Discuss that with him when he's ready to make contact with you," Albe said. "Other than that, I'll take that phone number from you when you're ready."

"He did tell me he could find out the answer to that for me . . . about an unauthorized phone call," she said. "That was made out of here about me."

"I didn't make a phone call," Albe said. "Is this something he can get back to you with?"

"Yes, yes, yes," she said before he shut the door again.

She sat on the bench and grabbed her knees. Slowly, she rocked back and forth, like a child who wanted to be held. After several minutes, she got up and stood in the corner by the door, trying to listen to what officers were discussing. She put her face in her hands. It was Monday afternoon, and she hadn't slept since Saturday night—and even then it was in fits and starts. She hadn't slept enough or eaten well in weeks.

The officer opened the door again.

"In case it's getting, in case it's going to go a long time, I'd like to make a list of the things—if he makes a phone call . . . in case the estimate is longer, and he needs to make a phone call for me, I would like to be able to write down a list of things that he needs to say," she said.

"He's going to make the call," Albe said. "Write down what you need him to ask before we give him a call, okay. Sound good?"

"If we run out of time . . ."

"He's going to make the call now if you want," Albe said "You want him to hold off on it?"

"I hadn't figured a cutoff time. I just wanted—where it's just going to take forever . . ."

"Okay, so you don't want to do it right now?"

"I would like to speak to you," she said.

"I'm going to hand you a pen and a piece of paper," Albe said. "Just think about what you want to add."

"Could I write the things down now? I just want to write down the things. Could you just write them down? I won't touch the pen or the piece of paper."

Albe closed the door and she finally broke down, softly crying, breathing heavily. She curled up on the bench and cried some more—her every move, her every word, videotaped inside the cell.

And then she walked the few steps to the door and sat on the floor, whimpering. She got up and stood in the corner again—trying to hear what was taking place on the other side of the door.

Like a lion kept in a cage too small, she paced back and forth—back in five steps, forth in five steps. She kept counting on her fingers and whispering something.

Then she started knocking on the door again. Knocking, knocking, knocking, over and over. Frantic knocking.

"I just want to have four words as reminders because I keep forgetting them . . . pick up, house, um um um, work, birthday cake," she said to the officer. "If I think of anything else, I'll tell you. Thank you."

Two officers closed the door, but she knocked again in a few minutes.

"I have one more . . . ARD—A-R-D—meeting," she said, referring to a NASA Aeronautics Research Directorate meeting.

She stood by the door again, listening. At that point, did she realize that her life as she knew it was over? Or did she think she would be able to walk out with no repercussions? She bent over, like a runner trying to catch their breath. Then she lay down again by the door and curled up in a fetal position, eavesdropping. Becton's office was just across the hallway.

After a while she knocked again, and Albe came to the door.

"May I have a water, please?"

He brought her water in a Styrofoam cup, and she slowly drank it.

"Let me know when you want him to make the phone call, all right?" Albe asked her.

"May I use the bathroom?"

Albe took her to the bathroom and then escorted her back to the holding cell.

Officer Greg Smith gave her an apple. "Try to eat that—you don't want to get sick . . . So you know where you are?"

"Orlando."

"You seem a little bit out of it," Smith said.

"I've been awake for a long time," she said.

"How long?"

"Since yesterday morning."

"You didn't sleep at all last night?"

"No—I've been with you."

She felt a little faint and paused.

"Feeling better?"

"No," she said.

"What's going on in your mind right now? What are you thinking?" Smith asked.

"That I may be locked up forever," she said, crying softly, "and that I think he liked me."

"I doubt very seriously you're going to be locked up forever," Smith tried to reassure her. "Did you think about what you were doing when you were doing it and how bad that looks, especially coming from your background?"

"I know the stuff looks bad," she said. "But I didn't, even if I wanted to, if I wanted to do something with any of those things, I couldn't have done that. I couldn't have something actually . . ."

"Why did you even have them?" Smith asked.

"At first I wanted her, I wanted her—I need to be able to talk to her," she said. "I'm afraid once I told her who I was and what I wanted to talk about, that she wouldn't talk to me and she might get me out of the car and hurt me or do something and I just wanted to make her stay and talk to me. I just wanted her to stay. You know—I, like, I didn't know what I needed to do that night and I brought all those different things and then I got there and I, I can't use any of these things. I can't even . . . I . . ."

"Calm down, sit down, and eat your apple, and I'll come talk to you in just a little bit," Smith said.

"If I sit down, will you talk to me?"

It was as though Smith was the only friend she had in the world at that moment.

"I just needed to talk to her . . . I didn't like while we were waiting for [the] bus."

"Couldn't you just call her?"

"I could not find her home phone number and I had a guess at a cell phone number, but the only reason I had that is because I saw a phone bill, and I just took a guess that that was the right number," she said. "I still don't know that it's the right number. I just saw a phone bill that was left out at somebody's house, and I just guessed based on how many times I saw the number that that was probably it. I only had a minute to look at it because I didn't want him to come back and find me."

"Where did you find the number—on his phone bill?"

"The person in question. Why was I there?"

"Are you and Bill dating?" Smith asked.

"Dating requires a public disclosure," she said.

"What do you think?"

"I can tell you we're . . ."

"Don't analyze it," he said. "Are you attracted to Bill?"

"I don't want to get . . ."

"I can tell by your eyes, you're analyzing everything I'm saying to you," Smith said. "Nothing you say is going to leave this room."

"I don't want to get anybody in trouble anymore," she said. "I'm already in trouble. If the only way, if this is going to come to something that's already worse than it's—if the only way I can explain what's going on is to make it worse for other people and . . ."

"You know what you need to think about right now is you," Smith said. "Your entire future depends on you right now. You know that."

"How could I do that?"

Someone outside the cell calls for Smith.

"You're not allowed to talk to me anymore?"

"You do the talking," he said. "I can't ask you any questions about what's going on. I just wanted to see what you were thinking. What's going on in your mind."

"I should've asked you first—You're not going to say anything to anybody, are you?" Nowak asked.

"No."

"Okay."

"You know if anybody else has?" Nowak had to know her status as an astronaut under arrest would make the news. But she had no idea that the story would go worldwide within twenty-four hours.

"No, nobody outside of this."

"I heard somebody making a phone call," she said. "I told him, I heard somebody making a phone call about it."

"The only people being contacted are police personnel. We're not calling anybody in the media or anything like that. We don't call the media for anything anyway," Smith said. "We let them find out about it."

"Somebody from where I work or associated with that?"

"Oh—that I don't know," Smith said. "I think by law we may have to."

"He already said that was not required," she pleaded.

"Oh, he did? Well then, he wouldn't."

"Somebody else might have."

"Not in this office. I don't know if they have or not."

Smith looked up at the camera.

And then Nowak treated him like a priest in a confessional.

"The time that was spent and the words that were shared in both directions were similar to the ones that were being shared elsewhere," she said, seemingly talking in circles. "I'm not in a position to be publicly shared. Does that make sense? I didn't know if she had the same understanding and I wanted to talk to her—that's all I wanted to do. That's it."

She cried softly again.

"I didn't have any plans. I just, I gathered things, not knowing what I was going to do with them or show them or any idea like that, but I certainly couldn't have used any of them and most of them are too scary to even think about, and that's why they were put away and the other one, I didn't even attempt to use that. I just wanted to sit there and talk to her.

Her words, bottled up for hours, spilled out in a torrent.

"And if she said she wouldn't talk to me and walked away, I just wanted to have something there to say please just stay here and just talk to me for a few minutes about this. But I didn't get to do that. Obviously, I didn't have a plan of doing anything harmful or even if I could plan to talk to her because I mean, I just walked up and I was trying to think of an excuse to where she would let me sit with her and talk with her, and that's stupid because I'm sure it looks weird to just have somebody come up and just ask to use your cell phone and give you a ride over by somewhere. Whatever. I just wanted to sit and talk with her. That's it."

"Well, I hope everything works out for you in the future and that you can get over this," Smith said.

She cried again.

"How long have you been in Florida?"

"Since yesterday . . . I didn't want to be gone very long. I just wanted to talk to her."

"Relax and have some water," Smith said, leaving the room.

A few minutes later when she knocked on the door again, Becton came to talk to her.

"Are you wanting me to make a phone call for you?" Becton asked. "I can't let you make it."

"He can dial it and then let you talk," another officer said.

"I'm sorry—it comes from above me," Becton told her. "But you'll be right there. It comes from above."

"But you'll have to say who you are and where you're calling from," Officer Rich Albe told her. "Is it your husband or ex-husband?"

"We're not living in the same place."

"We know that you're not living in the same house," Albe said. "Just communicate and stop analyzing. My guess is he probably already knows—through the chain of command. Is he in the military? Everybody has to be notified. You're in the Navy, correct? The victim may have called—I don't know if it was your commanding officer or her commanding officer, but he called and asked what was going on."

And so it was over—Richard knew.

"What did you think? That nobody was going to know?" Albe asked. "You're smart enough to know better than that. Or did you just want to have a chance to explain yourself? Were you planning on lying to them?"

"No—no. I'm not lying to anybody. Somebody else has . . ."

At that point, FBI agent Jose Lima walked into her cell, carrying her duffel bag.

"What's going on, I'm not going to get involved with," Lima told her. "I know you have a document in your car that belongs to NASA. Do you authorize me to get them and give them back to NASA?"

"There's nothing classified," she said.

"NASA OIG [Office of the Inspector General] requires me to get them back," he said.

"I guess I would request to go with you. Obviously, you can take what you want," she said. "There's nothing wrong with anything I have in my car."

"If it says mission or space shuttle, I'm taking it. If it's yours, you can get it back from OIG," Lima said.

She began to whimper.

"I'll come back, I'll sit with you there, and we'll go through it. Okay?"

Lima left the room, but Smith walked up into the open doorway.

"Even if everybody understands what happened. . . . I guess I didn't think there was anything to get caught," she said. After hours of going in circles with Becton, she was now talking to Smith.

"I just wanted to talk to her, and I just thought we'd talk and she'd understand why I was there and why I didn't want her . . . ," she said. "I thought that she knew—I was told that she knew who I was, and I wanted to, to see if . . . here's who I am and this is why I'm here to talk to you and I wanted to know if she had been dealt straight with and if not, then I thought somebody needed to be straight with her . . . and I didn't honestly know if I just wanted to know if he could be with somebody else. That was the only way to get a truthful answer. He made this great big deal about being honest with me . . . and not be honest about it. That's all right with me, but if it's not being applied everywhere, that's not. . . . it hurts. I thought this might be the only chance I'd have to talk to her."

"Other than on the phone?" Smith asked.

"She could hang up," Nowak said. "If that's what she intends to do, then there's no way to make sure we get to talk."

She saw all she had in her life—all she had ever had—up to this moment and saw all she was about to lose.

"It's not worth it, but I really, really believed it was," she said. "That was the whole point—I needed her not to walk away. I needed her not to disappear. I know it sounds stupid. I was trying to think of any way I could . . ."

"Don't try to explain anything to anybody but your superiors," Smith said. "You seem out of it—like you're not all here."

"It doesn't matter. The fact that I ended up here."

After a few minutes, Lima returned with a clear plastic bag filled with papers and folders. He began to take things out of the bag and, one by one, show her items.

"This is Bill's schedule, which he gave to me . . . This is a simulation schedule—there's nothing harmful in it . . . Personal bills . . . This is console—this is an updated draft procedure. Old draft procedures about robotics, these are my personal notes."

"Would you have a problem with OIG taking it?" Lima asked.

"They're personal notes and things that would be helpful to me," Nowak said.

"They're things the public shouldn't have—you follow me?"

They spent several minutes going through each folder. Then he packed up what he would turn over to the OIG, left the room a moment, then came back and took a second pile.

"I'll put these in the trunk of your car, okay?"

He left her alone.

Officer Andrew Welch walked to the doorway.

"Did a lawyer come?" she asked.

"Not here," Welch said. "Did you request one?"

"Yes. I was trying to be open and honest. I was willing to talk. I wasn't requiring that we wait for one, but I expected it. I did request it, and I made it clear that I requested it."

Smith appeared in the doorway again.

"Where's your children at right now?" Smith asked.

"He may have already picked them up. Or if not there, at the daycare center." Nowak said. "My son started . . . It was not my weekend to have them."

"So today was your pickup day? Any reason you didn't bring your cell phone?"

"I just forgot."

"Were you supposed to be at work today? When were you going to drive back?"

"Um, today."

"Who's going to be at your house tonight for your kids?"

"Sometimes he gets them if we make the arrangements."

In the distance, you can hear a phone ringing on speaker phone. She stands at the edge of the doorway, trying to lean out. After a minute, Becton walks up.

"He said he's taking care of the kids and he's already arranged a lawyer for you," Becton said. "The lawyer said no more official statements."

And with that, her conversation with the police officially ended.

23

★ ★

Take Two

On Tuesday morning, after a fitful night of tossing and turning, Becton arrived back in his office at the airport at 7:00 A.M. He plopped down in his chair and saw his desk phone's message light flashing—indicating he'd had some calls. He picked up the phone and was stunned that he had dozens of voicemail messages from reporters, colleagues, and supervisors. He fired up his computer and saw there were numerous emails, too, and he started to sort through them.

The day before, he had charged astronaut Lisa Nowak with four felonies:

attempted kidnapping;
attempted vehicle burglary with battery;
destruction of evidence; and
battery.

Then his phone rang. The caller ID indicated someone from the state attorney's office was on the line.

"Hey, this is Darlene Simmons." It was an intake assistant state attorney assigned to the Nowak case. She was also a reserve U.S. Navy judge advocate general. She told Becton she was concerned that Nowak's bond was set too low and then asked why Becton had not charged Nowak with attempted murder.

"I have probable cause to charge her, but all of the evidence is circumstantial, and there's no way to prove the attempted murder charge beyond a reasonable doubt," Becton explained.

"I would be able to prove the case with current case law to the standard of beyond a reasonable doubt," Simmons said. "That's our job. We make the [formal] charging decisions."

Knowing the magnitude of the case and the repercussions of making that charge with only circumstantial evidence, Becton didn't want to move forward without the advice of his supervisor. In his eighteen years of law enforcement experience, he had never before been contacted by anyone at the state attorney's office and told to add a more serious charge to a case.

"Let me put you on hold," Becton told Simmons.

Becton walked into the office of his sergeant, Dennis Ahern, and told him what Simmons wanted him to do. Sergeant Ahern was just as surprised as Becton. They put Simmons on speaker phone, and she told Ahern the same thing she had just told Becton. Ahern looked up at Becton, amazed that she wanted to add this charge.

The pair told Simmons they would get back with her and then got the Orlando Police Department legal adviser, Lee Ann Freeman, on the phone. Freeman said she was also surprised by this move and agreed with Becton that the evidence was only circumstantial and might not be proved in court beyond a doubt. But Freeman said to go ahead with the attempted murder charge if Simmons thought it could meet the legal standard.

Becton climbed into his squad car and drove over to the Orange County Jail on Vision Boulevard, with Interstate 4 in the distance. The parking lot was already filling up with large satellite trucks and smaller local live-shot vans. Becton was praying that he could walk into the building without being noticed by any local reporters who might know him. He walked down the sidewalk and didn't make eye contact with anyone, then entered the building, filled out the necessary paperwork to file the additional charge, then slipped back to his car, unnoticed for what would be the last time in several years.

24

★ ★

Media Circus

I had spent that Monday evening at home, alone, watching television on my tiny, fifteen-inch-screen TV, tin foil stuck to the antennae in the hope of getting better reception. After buying a house and then remodeling it following extensive damage from the 2004 hurricanes, my budget was tight and cable TV was out of the question, as was a new flat-screen or high-definition number.

I was watching the Orlando NBC affiliate—because that's what got the best reception—when the eleven o'clock news came on.

"An astronaut is under arrest tonight after police say she attacked a woman at an Orlando International Airport parking lot," sputtered the blonde anchorwoman, Martha Sugalski. It seemed like she couldn't get the words out fast enough. "Police say Lisa Nowak drove a thousand miles from Houston, Texas, to O-I-A wearing a diaper so she wouldn't have to make any stops. They say she went there to confront a romantic rival."

And that made me literally sit up and stare at the screen, eyebrows raised and mouth open in stunned silence.

Sugalski went on to report that Nowak had been part of the crew on the Fourth of July shuttle launch seven months earlier. Now, she had apparently pepper-sprayed a woman in Orlando International Airport's Blue Lot after trying to get into the car with her.

I picked up the phone and called a colleague. We liked to trade stories about what local tourism officials called the Space Coast, but we affectionately referred to as the "Skank Coast" because, beyond the space center and rocket scientists, were areas like Mims and Micco that were outposts for less savory characters journalists like to refer to as "Florida Man" and "Florida Woman."

"You're not going to believe this," I said, parroting back the pertinent facts.

"What!? You've got to be kidding!" reporter Billy Cox said.

"Nope. And get this. She wore a diaper."

Every reporter at *Florida Today* newspaper, where I worked, has taken the tour of Kennedy Space Center and knows that astronauts wear adult-sized diapers under their flight suits. They became standard issue after 1961, when Alan Shepard famously peed in his suit as he waited for hours to lift off in the Mercury "Freedom 7" rocket. But to wear them on any other occasion seemed absurd.

I covered breaking news at the paper, working the Tuesday-through-Saturday shift. I had spent Saturday covering a deadly tornado that had torn its way across central Florida, killing dozens of people and obliterating a small neighborhood of mobile homes near DeLand. I had hugged a woman whose mother's body was found curled in a fetal position amid the wreckage of what had been her home. The trailer's huge, heavy metal frame had wrapped around a pine tree. So Sunday and Monday had been days of physical and mental rest for me.

I went to bed that night wondering what would compel an astronaut, of all people, to behave in such a way. But I, better than most because of my job as a crime reporter, understood the lengths to which men can push women, or vice versa. I drifted off to sleep, trying to picture this woman I watched take off on July Fourth drive across five states to assault another woman.

My phone rang at 8:00 A.M. Tuesday morning.

"Hey there, Kimberlynn." It was my editor, Don Walker, an import from Shreveport, Louisiana, and a great newspaperman. He knew I had had a difficult time at the paper with a previous editor, so Don bent over backward to try to make my life a little easier. And I did the same for him—never turning down a request for help, a plea for filling an overlooked weekend spot, or an order to rush to a crime or accident scene.

"What's up?" I asked him.

"Hey, we've had an astronaut who was apparently arrested at Orlando International Airport early yesterday morning, and she's supposed to have her first appearance this morning. I don't think you'll be there in time for that, but she might bail out today. Maybe you can get her coming out."

"Yeah, I saw that last night on the news. Unbelievable."

"I know, I mean a frickin' astronaut of all things. Anyway, can you go over to the Orange County Jail?"

"Sure—when is she scheduled to get out?"

"We don't know—just sometime today."

He gave me directions to the jail and told me to meet photographer Craig Bailey and videographer Tim Walters. In 2007, newspapers were evolving into so much more than the paper document you read over a cup of coffee in the morning. They had websites for all the latest news and photographs, along with videos that viewers could download.

The drive from my home in South Brevard to the jail took me about an hour, as I sliced my way across the state on the newly renamed Beachline toll road, a desolate, four-lane highway that ran across a vestige of Florida's primitive landscape of sabal palms, maple trees, and orange broom sedge.

It was a Florida postcard day, with a cloudless blue sky and mild February temperatures—the kind of weather that lures northerners to move south to what had originally been a giant swamp. They realize their mistake in mid-August, when everyone in the state simply swelters in their own juices, fights off marauding hordes of mosquitoes, and boards up for looming hurricanes. But if you were born to it, as I was, there's simply no place else that really feels like home.

I knew I had found the jail when I saw the small nest of satellite and live-news trucks in the parking lot. And I knew more would come throughout the day.

In my then twenty years as a reporter, I had been at the center of the news universe three times before. I had been an anchorwoman and reporter in Israel during the Gulf War, standing on my balcony to watch Iraqi SCUD missiles rain down on Tel Aviv. I had worked for a wire service in Washington, D.C., when President Clinton was inaugurated. And I had been an anchorwoman and reporter in Tallahassee during November 2000, as the leadership of the free world was decided just a few miles from my tiny bungalow tucked into the north Florida woods.

I knew that today would be my fourth time at ground zero of international headlines.

I walked up and found photographer Craig Bailey, an infamous chatterbox with a good heart.

"Hey, what's going on?" I asked.

"Kim-ber-ly Moore! What is goin' ON?!" Bailey often talked in exclamation points and loved to pontificate. Reporters liked working with him because he was smart, nice, and, usually, pretty funny.

"Have I missed anything yet?"

"NOT. A. THING!"

I walked into the reception area of the jail, a plain, sterile room with fiberglass chairs by the window, and a large metal detector and security officers near the door. Off to the left, a woman sat behind a window, waiting to take bail money from family members, attorneys, or friends for the inmates inside. I recognized the jail's spokesman, Allen Moore, from the previous night's television news.

"Hey, I'm Kimberly Moore with *Florida Today* over in Melbourne," I said.

"Hey, Allen Moore," he said, shaking my hand. "I'll have something for you in a minute if you want to wait outside."

"Sure, thanks. She's already had her first appearance, right?"

"Yes. Now we're just waiting for her to be released."

I walked back out the doors and found a bench in a relatively shady area. Even though it was February, the sun could still blister my pale skin. Next to me, busily typing on a laptop, was a stocky man with glasses and a coat that was slightly too big.

"Hey, I'm Kimberly with *Florida Today*."

"Mike Schneider with the AP," he said.

"Are you filing now? From out here?"

He had his computer plugged into a wall outlet and an aircard was sending his story over the invisible tangle of cell-phone airwaves, back to the home office. Wi-Fi did not yet exist for the masses. Technology was making the news business a craft of speed but not necessarily precision. But what was wrong at 9:05 could be corrected online at 9:06.

"Yep," he said, hitting the SEND button. "Done."

"That's amazing. Do you know what's going on? What's the latest?"

Reporters often shared what little tidbits they knew, especially with out-of-town reporters—although they kept the best nuggets for themselves. They also knew that friendships forged at crime scenes and stakeouts could pay off later at another news event if they were the ones on foreign territory—or late.

"She's supposed to be released any minute," Schneider said.

We chatted for a few minutes about shuttle launches we had both covered and friends we might have in common.

After a while, the jail spokesman stepped out of the building and waited for the flock of media to circle around him. Reporters flew in from all angles—rushing up from the parking lot, from live and satellite trucks, and from a series of benches on a sidewalk alongside the building—to feed on the information. Cameras and microphones were everywhere.

Allen Moore announced that additional charges were being filed against Nowak. He didn't know what they were yet, but he would tell everyone as soon as he knew.

"We only know the charges are being filed as we speak," he said.

The spokesman also prepared the reporters to go upstairs for the hearing on those charges. No purses allowed, only pool cameras (which would take pictures and video to share with the others), and, of course, no weapons.

Among the gaggle of slightly disheveled print reporters and perfectly coiffed TV reporters outside the jail were two men who didn't fit in.

One was wearing blue jeans, a button-down oxford, and a brown leather flight jacket. The other was wearing jeans, a polo shirt, and a matching flight jacket. I approached them and introduced myself.

"Hey, I'm Kimberly Moore with *Florida Today* over in Melbourne."

"Steve Lindsey," he said.

"Are you with NASA?"

"Yes."

Lindsey was a retired U.S. Air Force colonel who had been Nowak's commander on her space shuttle mission. He was also in charge of the astronaut corps. He and his friend and colleague Chris Ferguson had flown from Houston to Orlando in a NASA T-38 training jet. Ferguson was there as the senior Naval officer in the astronaut corps.

"We're here representing NASA," Lindsey said, sticking to the talking points. "Our main concern is Lisa's health and well-being, and making sure that she's safe and we get her through this and we get her back to a safe place with her family."

Reporters strained to get closer to this blond man with a thin build. Lindsey was an average height—the perfect height for a pilot to sit in a cramped cockpit.

"This is a private . . . it is a personal matter; it's a legal matter that she and her family have to deal with."

"What's your function here today?" someone asked.

"We're down here supporting her like we would any employee at NASA, if they were in this situation," Lindsey answered. "We're a close family, and we try to take care of our own."

"How is she doing—have you talked to her?" another reporter asked.

"I can't comment on her status—that's an ongoing thing. I know her well. I've flown with her. She did a fantastic job on the mission. She's been a great asset all the years that I've known her."

"What did she do on your flight?" I asked.

"She operated the robotic arm and was a flight engineer."

A little later, Nowak's attorney, Donald Lykkebak, came out of the jail. He was an imposing figure with a furrowed brow, and he showed a palpable contempt for the press corps. He looked a little like Perry Mason—without the charm. He wore a dark-blue suit, red tie, and a U.S. Marine Corps lapel pin. Over the next two and a half years of court appearances, I would come to know these clothes as his standard uniform.

The book *Lawyers of Orange County* by Buckie Allen described Lykkebak as a native of Illinois who wound up a glee club singer his freshman year of college and toured Europe the next summer with his fellow entertainers. He worked in the Judge Advocate General Division in the Marines, but it was his work as a prosecutor in the felony division with the Office of the State Attorney, Ninth Judicial Circuit, that gained him the nickname "Captain Ice." In 1977, he opened his own practice.

"The facts recited in the charging affidavit are insufficient to argue a case of attempted kidnapping," Lykkebak barked at us. "What it does allege factually is a case of battery and maybe, with a stretch, this attempted vehicle burglary with battery."

Lykkebak went back into the jail. The media updated their stories and then went back to waiting. Then waited some more.

The jail spokesman came back outside, and the pack gathered around him once again. He went on to explain what was happening.

"Just moments before the anticipated release, the Orlando Police Department came into our booking center to file an additional charge of attempted first-degree murder, and that stopped the release process from going forward," he said. "The next step is for the court to hear on this charge and have an initial appearance on that."

He explained that Nowak was in protective custody—meaning she wasn't milling about with regular killers, thieves, and thugs.

"She's been by herself, and the cell is under observation," he said.

After he spoke, I flipped open my cell phone and dialed Don Walker's extension as fast as I could.

"They're charging her with attempted murder," I told my editor.

"What?!"

"Yep—let me dictate something to you."

I could hear the computer keys rapidly pecked as he typed the breaking news story with two fingers. A few minutes later, it was the main headline on the paper's webpage, floridatoday.com.

Becton had filed a new affidavit on the attempted murder charge. It read:

> The facts that Mrs. Nowak drove approximately 900 miles, urinated in diapers so that she did not need to stop, stayed at a hotel where she paid cash and used a false name and address to register, stealthily followed the victim while in disguise, and possessed multiple deadly weapons at the time she confronted the victim, as well as spraying the victim with a substance meant to disable a person, create a well-founded fear and give this investigator probable cause to believe that Mrs. Nowak intended to murder Ms. Shipman.

I walked over to my dark-blue Mustang and opened the trunk. Inside, I had a small cooler with a couple of sandwiches I had bought at a grocery store in Deland, near the weekend's tornado damage. The cool winter temperatures had kept them edible, and I offered one to the paper's videographer, Tim Walters, who happily accepted. Anyone who thinks journalism is glamorous has never eaten a three-day-old sandwich on a hard steel bench outside of a jail.

Among the gaggle of reporters, cameramen, live and satellite truck operators was what looked like a family bailing someone out of jail. Like the reporters, they had been there for hours. It appeared to be a mother, possibly a daughter and son in their twenties. They were all dressed in clothing that seemed to be from a discount store's clearance rack, and they kept going in and out of the lobby. I would find out why they were there at the end of the day.

Another two hours went by before the pack of reporters were summoned. Finally, sometime after noon, Spokesman Moore called us into the lobby. One by one, we waited to go through security and be scanned for hidden weapons. Moore escorted us upstairs, where we found ourselves in a fluorescent-light-filled hallway, with a small viewing area just around the corner. Peering through Plexiglas, the reporters looked into a bare-bones courtroom. The austere, stately, wood-paneled courtrooms of TV dramas would weep at the sight of this—plain, whitewashed concrete-block walls, a dais for the judge, and two small podiums for defense and prosecuting attorneys.

I sat down in the front row of the observation room, and Lindsey and Ferguson took a seat behind me. A *Washington Post* reporter sat down next to me, fresh off a flight into town. I got him up to speed on the case as we waited—and he diligently scribbled notes, including quotes.

I looked up to see stalwart *Orlando Sentinel* still photographer Red Huber adjusting his lens.

"Hey, Red!"

"Well, hey, it's good to see you again." We pecked each other's cheeks.

We had gotten to know each other at a Delta rocket event at Cape Canaveral Air Force Station about a year before. That day, the public information officer had corralled us onto a bus, driven us out to the launchpad, and then apologized as we waited, sitting on the pavement, for the rocket to rise into an upright position. The crew, it seemed, had gone on break and wasn't willing to accommodate the media's schedule. But it gave me at least an hour and a half to get to know Huber, who had worked at the *Sentinel* for decades.

"How are things going for you these days?"

"Pretty good. Better, actually," I said.

As we spoke, Nowak entered the courtroom. It was the second time the press had seen her since her arrest, and it was the first time I had ever seen her in person. She was tiny—it seemed her shadow might weigh more than her body. Her legs were shackled at her ankles, and her hands were cuffed to a chain around her waist. Her long, wavy hair trailed down her back.

She was wearing a dark-blue jail jumpsuit. It was a substantial demotion from the bright-orange "pumpkin" astronaut flight suit she had worn just seven months earlier.

And she looked gaunt. Her NASA photo showed a bright, healthy woman, with glowing skin, round cheeks, and a warm smile. But this woman had sunken cheeks and dark circles under her eyes. I thought she looked confused and a little frightened as she stood in front of Circuit Court Judge Mike Murphy, with Lykkebak at her side.

After some formalities, Murphy told Nowak that she would not be allowed to speak with Colleen Shipman and she would have to wear an ankle monitor that would track her every move.

"You wouldn't even be allowed to send flowers to say, 'I'm sorry,'" Murphy said. "Do you understand?"

Barely audible, she said, "Yes, sir."

A clearly angry Lykkebak spoke up, pleading with the judge to throw out the attempted murder charge.

"The police department is seeking a second bite of the apple today," he said.

Prosecutor Amanda Cowan shot back: "A review of all the circumstances—printing out a map, driving from Texas, using a disguise—all of the facts upon further investigation led to that additional charge."

"This is based on the same circumstances—are there additional facts in this?" the judge asked of the charge.

"She was arrested at 4:00 A.M., they held her at a facility until 6:00 or 7:00 P.M. last night. They had custody of her for fourteen or fifteen hours, during which time they subjected her to continual interrogation," Lykkebak nearly shouted. "They had all this information yesterday when they filed the initial charges."

Lykkebak also told the judge that Nowak was kept from him for hours, even though he ordered police to stop questioning her.

"What she did was spray her with pepper spray and no more," Lykkebak said, astoundingly admitting guilt to one of the misdemeanor charges. "What she said was not that she would kill her, but that she wanted to talk to her."

The judge asked if Lykkebak thought the evidence was proof enough.

"What we have is a desperate woman who wants to have a conversation with another woman. She doesn't shoot her. She doesn't stab her. She doesn't do anything except spray her with pepper spray. She just wants to talk to her. I don't follow where we can keep jacking this case up every three hours."

Amanda Cowan moved for no bond—meaning Nowak would be held in jail until her trial.

"I will find probable cause," Judge Murphy said. "I'm not going to find that the state has proved the guilt is great."

Murphy felt there was enough to move forward with the charge, but cautioned that there would need to be more evidence at trial.

"Set aside your awareness that she is an astronaut," prosecutor Cowan said. "I would ask that you would consider her as any other defendant and leave her bond at none. She had a mission that she was very determined to carry out. I think but for the victim's awareness of her surroundings and quickness to get in her car, this would've had a different outcome."

"When a person stands before a magistrate such as yourself, his or her past record or history are exceptionally important," Lykkebak pleaded. "If your good work doesn't come to help you now, when does it?"

The judge ordered her released on $10,000 bond on that count, in addition to $15,500 for the other charges of attempted kidnapping, burglary of a conveyance with battery, and misdemeanor battery. If she were convicted, she would face a possible life sentence.

A little while later, all the reporters went back downstairs and sat and waited for the broken hero to leave the Orange County Jail. Her family had posted her bond.

I called in an update to Don, then walked into and out of and back into the lobby. I joined the AP's Schneider on one of the fiberglass chairs inside. Other reporters started to swarm in the lobby. It was late afternoon, and the sun was starting to sink and glare in the windows.

Astronauts Steve Lindsey and Chris Ferguson sat across the room in chairs by the door out of which Nowak would walk into the lobby. They looked like goldfish spilled onto a tabletop.

A "perp walk," as they're called by cops and reporters, involves a crime suspect walking into or out of a jail, usually escorted by an attorney, bail bondsman, or law enforcement to or from a car. They usually take less than a minute, but they can be traumatic for someone not used to the criminal justice system.

I had worked in public relations for eighteen months at Habitat for Humanity International, handling events with U.S. presidents, NFL stars, and multiple homeowners. I had literally written the book for Habitat on crisis communications—what to do when something goes horribly wrong.

I wanted to walk over to them and say, "Hey, here's the deal. She's going to walk out those doors and walk to your car. That's going to take, what, something like twenty seconds. Granted, it's going to be a long twenty seconds, but when she's in the car, it's over. She doesn't have to say anything to us. Just hold her head up and walk to the car."

But I couldn't cross that line. I wasn't in PR anymore, and it would have been absolutely taboo for me to give them advice. And besides, surely, NASA PR folks were telling them what to do.

Finally, the jail spokesman asked all the reporters to leave the lobby and wait outside. The windows were tinted, so we couldn't see in. But those inside could see out. I planted myself right beside the door. Mine would be the first face Lisa Nowak would see when she walked out.

As the sun set orange over the line of media trucks, their poles and dishes reaching skyward, we could see someone walking toward the door. A hush fell over us as we waited to see if it were her. This was what we had been working for all day.

And then the door opened, and I saw her and quietly uttered, "Oh, no."

Nowak walked out, her head covered by a gray jacket. It was, as I call it, "the veil of the guilty," used by many perpetrators as they walked in front of media cameras, as though we wouldn't be able to see them.

Nowak's bail bondsman, the woman appointed to put the ankle monitor on her, and the two astronauts guided her through the mob of nearly three dozen journalists.

Everyone glommed around her—she was the nucleus in this atom of information gathering. No one said anything at first.

I wanted no part of this mob scene, so I held back, walking about ten feet behind everyone else. Clearly Nowak wasn't going to make a statement to reporters, so there was no point in being a part of the pack of cackling hyenas waiting to tear into their prey. Craig Bailey would get his picture, and Tim Walters would shoot some video that we would post on our website.

And then the shouting started—not from a journalist, but from the family that had been hanging around the jail complex all day—mother, daughter, and son.

"WHY'D YOU DO IT?!" the mother shouted. "WHY'D YOU TRY TO KILL HER?!"

This self-appointed Greek chorus moved alongside the reporters, howling their questions.

"WHY DID YOU DO IT?" one of them shouted again. "YOU TRIED TO KILL HER!"

What the hell is this? I thought. *Audience participation?* I had never seen anything like it.

I watched as Lindsey maneuvered to the right of the mob, trying to make his way to the driver's side of the car. A cameraman who didn't see him elbowed him in the head. This space shuttle commander, accustomed to quasi-worship by the press in very controlled environments, had just been inadvertently hit.

Dave McDaniel, with WKMG, finally asked, "Captain Nowak, you have to say SOMETHING."

No she doesn't, I thought. *As a matter of fact, she shouldn't,* and I was sure her attorney had told her as much.

The group got into the car, and Walters lay on the hood of the maroon sedan to get a shot of Nowak inside. After he stood up, the maroon car slowly backed up and drove off, the sea finally parting for her.

Walters, just a few years out of college, walked up, video camera in his hand. "We waited all day for *that?*"

I slowly nodded.

Across the street, in the bail bondsman's office, an ankle monitor was placed on Nowak's leg that would track her movements—and that would call Shipman's cell phone if Nowak entered Brevard County.

Out in the Blue Lot at Orlando International Airport, WESH-TV reporter Greg Fox and cameraman Frank Burt were setting up to go live at 6:00 P.M. Burt looked down and noticed a slim, black can in the parking lot, behind a bus stop. He rolled it over with his foot and saw that it was a can of pepper spray. Fox called the police, who retrieved the can to dust it for fingerprints.

25

★ ★

Center of Attention

The day of the attack, Billy Oefelein flew from Houston to Florida to comfort Shipman at her Cape Canaveral townhome, and he remained with her for about three weeks. The next evening, he submitted to questioning by the Orlando Police Department (OPD) and the NASA Office of the Inspector General (OIG) in the astronaut crew quarters at Kennedy Space Center. Becton, OPD detective Michael Moreschi, and NASA OIG special agent Wade Krieger met with Oefelein at about 8:00 P.M. in the crew quarters meeting room, which included some mid-century-modern chairs and a sofa, along with a well-stocked bar. They placed him under oath before beginning their questions.

Oefelein, wearing regular street clothes, explained to the detectives that he had been in an exclusive relationship with Shipman for about a month and had been dating her for three months, beginning in early November. He said he met Nowak when both were stationed with the same squadron in 1996, and they became friends as they did rendezvous training together at Johnson Space Center, during which they learned how to dock the space shuttle with the International Space Station. They also went on winter survival training together in Canada in January 2004.

Becton said interviewing Oefelein was a lot like interviewing Nowak.

"It was obvious he was withholding his emotions during the interview," Becton said. "He was very careful of how he answered the questions presented to him. I guess we were about five minutes into the interview, and it was obvious he was caught between a rock and a hard spot. He knew that he was going to be in trouble with the military for his involvement with Lisa."

When asked how long ago he had divorced his wife, Michaella, he responded "two years ago"—in 2005. And when asked when his relationship with Nowak became romantic, he replied "three years ago"—in 2004.

Becton assured him that he was not going to Naval Criminal Investigative Services with information he gave until Becton completed his investigation.

"Once I addressed this issue, he relaxed a little," Becton said. "It was almost like he realized it is what it is, and he would have to face the consequences at some point. I could at least give him some time to get things in order."

And so Oefelein opened up more about the relationship.

"I would say that we were somewhat exclusive," Oefelein told the investigators about his relationship with Nowak. "Nobody prohibited anything, but I would consider it was exclusive for a period of time."

Becton thought Oefelein seemed respectful, genuine, and sincere, but very stressed.

"It was obvious to me that his concern was about his newfound love, Colleen," Becton said later. "I think if I had offered to take him to Colleen, he would've jumped in my car in a single bound."

Oefelein told the investigators that during the second week in January 2007, he had explained to Nowak that he had met Shipman and he wanted to pursue a serious relationship with this new woman. He said Nowak seemed disappointed but accepted the news, adding that their relationship had not been like it was before he had met Shipman because he wasn't calling Nowak as often anymore, even though Nowak was calling him a lot.

"I wasn't always receptive to the phone call," Oefelein said, adding that Nowak would leave friendly messages when she called. "Not adversarial, just indicative of the type of person she is—just a friendly person."

He said he had not cut off their friendship and that they still trained together at the space center gym and on bicycles for a competitive bike race in April.

Detectives were trying to figure out how Nowak obtained Shipman's flight itinerary, asking Oefelein if he locked his computer at work, where Nowak would have access to his desktop. He said he let the computer "go to sleep" after five or ten minutes of nonuse and it would require a password to log back in. They also asked if Nowak had a key to his apartment—he said she did.

"As far as I know, she still has it," he said.

He told them that he had a photo of Shipman on his desk at work—which Nowak would have access to see—and he also had some photos on his hard drive at home of him and Shipman from the party the night they met.

When detectives asked if Nowak had been a nice girlfriend, Oefelein gave an answer that seemed to contradict his earlier statement about them having an exclusive relationship.

"It's hard to consider her a girlfriend, she was an ex-interest," Oefelein said. "We had a relationship, but, you know, never really said the word 'girlfriend . . . ' I would consider Lisa to be one of my best friends at NASA." He thought she felt the same about him.

Oefelein then said that Lisa probably had not confided in anyone about their relationship because she was an intensely private person.

"I don't know—maybe that's part of the problem, as an amateur [psychiatrist]," he added. "I would have never predicted this. Her behavior indicated to me that she was stable and accepting of what I'd told her with Colleen. She had actually wished me a nice weekend, knowing Colleen was gonna spend it with me."

Detectives asked if he had ever seen any anger issues in Nowak.

"She seemed extremely level-headed and nonemotional. She didn't flare up, she didn't have a temper," Oefelein said. "I'm trying to remember ever seeing her even angry, and I don't [think I've] ever seen her even angry, you know, the years I've known her. I mean, she just seemed to be very quiet, somewhat shy, a very private person."

After talking with Oefelein for about twenty minutes, the interview was over. It was the first and last time Oefelein has ever spoken publicly about Lisa Nowak or the events of February 5, 2007.

"I had the impression he was afraid that Lisa had screwed up his entire life," Becton said.

26

★ ★

The Fallout

The fallout from the Nowak incident was both swift and excruciatingly embarrassing for her, her family, and NASA. Comedians had a field day with the entire episode, especially the diapers, and it provided fodder for their jokes for months.

"The number-one sign an astronaut is trying to kill you," CBS *Late Show* host David Letterman quipped, "she keeps stabbing you with a pen that writes upside-down."

Members of her family flew to Houston to be with her after she flew back home and after she was evaluated by NASA doctors.

"We are naturally saddened and extremely concerned about the serious allegations being made against Lisa. We love her very much, and right now, our primary focus is on her health and well-being," her family said in a prepared statement. "Considering both her personal and professional life, these alleged events are completely out of character and have come as a tremendous shock to our family. We are anxious to allow the facts to develop so that we can better understand what happened and why."

They talked about her accomplishments and her "unblemished record" in the military and at NASA.

"Personally, Lisa is an extremely caring and dedicated mother to her three children. She has been married for 19 years, although she and her husband had separated a few weeks ago," they said. "We hope that the public will keep an open mind about what the facts will eventually show and that the legal system will be allowed to run its course. Finally, we are very grateful for the expressions of love and support that we have received from family and friends, and we ask for your continued thoughts and prayers for our family."

Her sister, Andrea Rose, told *People* magazine that Nowak never fully recovered from the loss of her friends on board *Columbia*.

"We knew Lisa was under a lot of stress," Andrea Rose told *People*. "But there is no way of knowing how a particular person will react to stress. We love Lisa and we're worried about her well-being."

NASA took the incident very seriously. When Nowak's commercial flight touched down in Houston, a NASA car was allowed to pick up her and her space shuttle commander, Steve Lindsey, on the tarmac, and she was driven straight to Johnson Space Center for a medical evaluation. She was also immediately relieved of her duties as the lead CAPCOM officer for the next shuttle mission, scheduled for the following month.

In addition, NASA emails show that both she and Oefelein were barred from accessing computers at NASA.

That day, NASA administrator Michael Griffin ordered a review of the agency's psychological and medical screening and care of astronauts.

Multiple astronauts and support personnel were interviewed by investigators in the days and weeks that followed. Two things were said several times by her colleagues about Nowak: she was a good and devoted mother and she was a hardworking astronaut who showed attention to detail.

One person who declined to be interviewed by investigators: Richard Nowak, Lisa's husband. He did, however, cooperate with Dr. Richard Pesikoff, one of Lisa's psychiatrists.

In July 2007, the NASA review committee issued a damning report, which concluded that supervisors would not listen to subordinates regarding astronauts' level of psychological readiness for space; related anecdotal testimony that astronauts launched into space inebriated; and recommended that NASA establish a code of conduct for astronauts. Finally, a congressional hearing was held in September 2007.

NASA committee members reviewed documents and interviewed NASA staff, including all behavioral health providers, all clinic-assigned flight surgeons, and eight of twenty-one space medicine division flight surgeons. Fourteen astronauts were interviewed, along with five of their family members. The committee's findings were far-reaching.

NASA's psychological testing evaluation has long been utilized to identify applicants who can readily adapt and perform effectively in the extreme environment of spaceflight. But, committee members found that the results of the evaluations are rarely and inconsistently used. Details of methods and criteria used, and data collected (all vital to even the most basic psychological evaluation) either did not exist or were not made available to the committee for review. The panel recommended that NASA charter an expert panel to determine "what, if any," psychological testing should be performed and how it should be used to select astronaut candidates suitable for space operations. NASA subsequently adopted a policy of enhancing the use of behavioral health data in the astronaut selection process.

The committee also identified an issue with the consistency of astronaut care. When most people seek medical help, they see their primary-care physician, and that doctor is usually the same one they see consistently over time. When the astronauts seek medical care at the flight medicine clinic at Johnson Space Center, they will see any one of a number of flight surgeons assigned to the clinic. And while a crew typically has one flight surgeon assigned to it, the 2007 report states "that relationship normally does not predate the assignment and does not persist beyond the immediate post-flight period," despite the fact that research shows continuity of care over time with one doctor "increases the quality of medical care and increases the detection of behavioral and psychosocial issues."

Also alarming was the fact there is no periodic psychological evaluation or testing conducted on astronauts.

"Once selected as an astronaut candidate, astronauts have no psychological evaluation for the remainder of their careers unless selected for long duration missions," the report stated. "There is no routine behavioral health assessment for commonly occurring issues such as depression, anxiety, relationship stress, substance use, or the cumulative effects of normal life events, all of which can lead to a decrement in performance."

But, the committee added, an initial screening and recurrent psychological evaluation are not intended to, nor can they, predict a future disorder of conduct or "act of passion." However, they can identify people at increased risk, allowing an intervention that might prevent an incident like Nowak's.

In response to the committee's recommendation, in 2008 NASA adopted a policy that behavioral health evaluations be integrated into the annual flight physical for all astronauts. These evaluations should be conducted by the flight surgeon responsible for the continuity of the astronaut's care, in consultation with behavioral health. NASA also adopted the committee's recommendation to provide regular training regarding behavioral health to flight surgeons.

No one interviewed for the report had any knowledge of an astronaut seeking behavioral health care. Patricia Santy, a former NASA psychiatrist and flight surgeon, said one reason for this lack was due to the fact that a bad psychological evaluation based on the subjective opinion of a psychologist could ground them. The whole point of their being at NASA was to fly into space, so protection of privacy was an area of paramount concern to the behavioral health providers, astronauts, and family members.

"This prevents effective communication regarding patient status between behavioral health and other providers, and severely limits the ability of the flight surgeons to make appropriate aeromedical dispositions," the report stated.

As a matter of fact, prior to Nowak's incident, astronauts were not required to report illnesses, injuries, or medication use unless they—the astronaut, not a doctor—determined the issue to be significant. This was a major deviation from what's known as the NASA Personnel Reliability Program (PRP), similar to those administered in the military, where every episode of illness, injury, or medication must be evaluated by a provider qualified to make fitness-for-duty determinations. The committee recommended that astronauts get on board with the PRP program.

The committee also concluded that NASA needed to adopt a formal "Astronaut Code of Conduct," behavioral and moral guidelines to which the group should adhere. NASA implemented such a code in 2007.

The code states that it "calls for a constant commitment to honorable behavior and ensures the continued privilege of participation in our Nation's space program." Its basic tenets call for:

Competence—"We recognize that lives are always at risk in human space flight and make extra effort to know our jobs and to perform them flawlessly."

Teamwork—"We are highly visible members of a large team of skilled and dedicated people. Mission accomplishments belong to the team, not to the individual."

Integrity—"We will strive to avoid the appearance of impropriety and readily acknowledge the limits of our expertise."

Relationships—"We will protect and balance the best interests of our co-workers, families, and NASA."

Personal Behavior—"We will live in a manner that reflects credit and honor upon our profession. Astronauts will respect and uphold the law and their duties as citizens."

Stewardship—"We are custodians of valuable national resources and will avoid conflicts of interest that could lead to their exploitation."

Lifelong Commitment—"Our responsibilities to each other continue after our time at NASA. We will always consider the implications that our actions have on all active and former Astronauts."

They also recommended that there be clear leadership roles in the astronaut office "to provide enduring supervisory relationships that extend over years and are not limited to technical or mission assignments. Supervisors should be senior astronauts, and each should have a manageable number of astronauts to supervise. These supervisors should report to the Chief of the Astronaut Office." Their roles would include performing twice-yearly performance evaluations of astronauts.

"While cultural changes are the most difficult to achieve, they are also the most significant and pose the highest risk of human failure if not adequately addressed," the report concluded.

One of the most damning findings of the report was that interviews with both flight surgeons and astronauts identified episodes of heavy alcohol use by astronauts in the "immediate preflight period, which has led to flight safety concerns."

"Two specific instances were described where astronauts had been so intoxicated prior to flight that flight surgeons and/or fellow astronauts raised concerns to local on-scene leadership regarding flight safety," the report stated. "However, the individuals were still permitted to fly . . . Many anecdotes were related that involved risky behaviors by astronauts that were well known to the other astronauts and no apparent action was taken. Peers and staff fear ostracism if they identify their own or others' problems."

In addition, incidents were described in which major crew medical or behavioral problems were identified to astronaut leadership and the medical advice was disregarded. This disregard was described as "demoralizing" to the point where they said they are less likely to report concerns of performance decrement."

NASA deputy administrator Shana Dale announced that NASA's

existing alcohol-use policy with the agency's T-38 training jets had historically been applied to spaceflight and was now explicitly extended as an interim policy to flight on any spacecraft, prohibiting alcohol use for twelve hours prior to flight and further stating that astronauts will neither be under the influence nor the effects of alcohol at the time of launch.

The panel recommended that NASA institute specific policies, procedures, educational efforts, and disciplinary actions to foster a culture that holds individuals and supervisors accountable for safe and responsible use of alcohol. That included, but was not limited to, a mandatory alcohol-free time period prior to flight.

As for Nowak, one month after her arrest, NASA issued a statement: "U.S. Navy Capt. Lisa Nowak's detail as a NASA astronaut has been terminated, effective March 8, by mutual agreement between NASA and the U.S. Navy."

"I called her last week, and she sounded pretty depressed to me," shuttle pilot Mark Kelly, her crewmate on STS-121, told *Texas Monthly* magazine three weeks after her arrest. "We are going to try to get together with her this week. She has nothing to do. She just sits in her house all day, every day."

The Navy took her back, and she was assigned duties at Corpus Christi Naval Air Station while she awaited her criminal trial and Navy hearing.

According to Nowak's testimony during her August 2010 Naval hearing, the chief of staff in Corpus Christi welcomed her to the base, saying, "You're on our team now."

The admiral told her, "We value your experience as both a naval aviator, having gone through the program that we're working on here, and flight experience you've had." Then he added, "We're proud to have you on board and we want you to do the best job you can."

Two and half months later, NASA also released Oefelein back to the U.S. Navy.

"The Navy and NASA mutually agreed to end his detail to NASA," said Kylie Clem, a spokeswoman at Johnson Space Center in Houston. "NASA has determined that his detail is no longer required for the purposes for which it was originally granted."

Oefelein and Nowak were the first astronauts to be sent back to their military branch following a public scandal.

The internal review and report, along with the congressional hearing, led in 2008 to an emphasis on the behavioral health and performance element of astronauts and the requirement of an annual psychological screening for astronauts.

"Before 2008, psychiatric screening was only done for astronaut candidates," Nick Kanas, a psychiatrist who once screened NASA astronauts, told *Discover* magazine in 2018. "Now all astronauts are screened by a psychiatrist every year regardless if they are flying or not."

The *Discover* article stated: "BHP offered space station astronauts training on conflict management, cultural sensitivity and coming to terms with the isolation of space," said James Picano, senior operational psychologist with NASA's Behavioral Health and Performance group.

"Some of the changes the group made include allowing each crewmember to have weekly video conferences with family members and opportunities for counsel with the BHP team. After the mission, crewmembers take part in a reintegration program, Picano told *Discover*."

27

★ ★

The Right Psychological Stuff

America's space program germinated in the years between the wars in Korea and Vietnam, when jet fighters had become commonplace and commercial jet travel took off. The National Aeronautics and Space Administration (NASA) was established in 1958 to counter the Soviet Union's launch of the first-ever satellite around Earth, *Sputnik*. Both countries then scrambled to put the first man into space, and NASA utilized both American and Nazi Germany rocket technology.

Psychologically speaking, most astronauts chosen for the Mercury program and those chosen today share several traits. According to a 1959 article in the *American Journal of Psychiatry*, Captain Edwin Levy, an associate investigator for the Mercury project, and Dr. George Ruff pointed out the commonalities:

> These were comfortable, mature, well integrated individuals. Ratings on all categories of the system used consistently fell in the top third of the scale. Reality testing, adaptability and drive were particularly high. Little evidence was found of unresolved conflict sufficiently serious to interfere with functioning. Suggestions of overt anxiety were rare. Defenses were effective, tending to be obsessive compulsive, but not to an exaggerated degree. Most were direct, action-oriented individuals who spent little time introspecting. . . . [M]ost showed the capacity to relate to others. They do not become over-involved with others, although relationships with their families are warm and stable. . . . [M]ost of the final 31 men had made excellent school and social adjustments. Many had been class presidents or showed other evidence of leadership. . . . [They had a] conviction that accidents could be avoided

by knowledge and caution. They believe that risks are minimized by thorough planning and conservatism. Very few fit the popular concept of the daredevil test pilot.

Former NASA psychiatrist and flight surgeon Patricia Santy wrote in her 1994 book *Choosing the Right Stuff: The Psychological Selection of Astronauts and Cosmonauts:*

Astronauts must also live with similar paradoxical limitations that continually stress their adaptation abilities. They must be sure of themselves, but not too sure—competent, but not arrogant; they must love to fly, but appropriately fear flying; they must be confident about their abilities, but not grandiose; they must react quickly to danger, but not be impulsive; they must obey authority, and yet be independent; and so on. Frequently, the bottom-line behavior exhibited by those who had the right stuff was narcissism, arrogance and interpersonal insensitivity.

According to Santy, the "Gemini and Apollo Astronaut Psychological Selection Criteria"—famously referred to by author Tom Wolfe in his book about the early space program as "the right stuff"—included:

1. General Emotional Stability: the absence of neurotic or psychotic symptoms and freedom from problems in the social, marital and financial spheres, as well as the ability to tolerate stress and frustration without emotional symptomology or impaired performance.
2. High Motivation and Energy Level: the demonstrated ability of the candidate to pursue realistic and mature goals with determination and initiative and the capacity to think in a creative, flexible manner when unforeseen events occur.
3. Adequate Self Systems: a clinical term suggesting a high self-confidence and the capacity to give opinions and make independent decisions, with the ability to depend on the judgment of others when the mission warrants.
4. Satisfactory and Productive Interpersonal Relationships: The ability to form satisfactory and productive relationships with supervisors, peers, and subordinates and to function as a team member in any role without being overly dependent on people for satisfaction.

The 2016 NASA report *Evidence Report: Risk of Adverse Cognitive or Behavioral Conditions and Psychiatric Disorders* states there are several "'trait characteristics' that reveal the 'right stuff,' the 'wrong stuff,' and 'no stuff.'" They haven't changed much in the sixty years since NASA's formation, and they include:

Instrumentality—goal-seeking and achievement orientation.

Expressivity—social competence in interpersonal relationships.

The "Big Five" factors of personality—NASA is looking for people who are not neurotic, but are extroverted, open to experience, agreeable and conscientious.

Resilient—patterns of positive adaptation in the face of significant adversity or risk . . . endurance when faced with unremitting stressors and bouncing back from stressful events.

Hardy—have a strong commitment to their values, goals and capabilities, a greater sense of control over what happens in their lives, and a perception of stressors as challenges to be mastered rather than events to be endured.

Neil Armstrong was the embodiment of all of these.

His resiliency served him well aboard Gemini 8 in 1966, when one of the earliest low-Earth-orbit missions suffered a critical system failure when a thruster stuck in place, sending the capsule spinning uncontrollably. In a matter of moments, Armstrong recognized what the problem was, came up with the solution, and implemented it—saving his life and that of pilot Dave Scott.

"This guy was brilliant. He knew the system so well," Scott told PBS's *Nova* in 2014. "It was my lucky day to be flying with him."

On May 6, 1968, Armstrong famously ejected from a lunar landing training craft just before it crashed and exploded, walking away with only a bit tongue. He was later found in his office, doing paperwork after the near-death experience.

"There was work to be done," he would later shrug and tell various interviewers in his midwestern, modest, and humble style.

Finally, when Apollo 11 was approaching the planned site of the first moon landing, Armstrong realized it was not a good choice because of boulders and a crater. He continued flying until he found a safer spot—coolly setting down with only about forty-five seconds of fuel left.

Multiple experts—from Apollo 11 astronaut and moonwalker Buzz

Aldrin, to Dr. Jonathan Clark, a former NASA doctor and the widower of space shuttle *Columbia* astronaut Laurel Clark, and Dr. Santy—agree: NASA has achieved the greatest technical feat in human history by landing men on the moon, along with building and maintaining the International Space Station (ISS). But they do poorly at psychologically supporting astronauts.

Aldrin, the second man to walk on the moon, has honestly and bravely discussed his own battle with depression in several books. He wrote in his 2009 autobiography, *Magnificent Desolation: The Long Journey Home from the Moon,* that in 1969, his "life was unraveling all around" him following his lunar walk. Not only was he dealing with "the melancholy of things done," he was also still mourning the death of his mother, Marion Moon, who committed suicide the year before his moonwalk. He has said she could not handle the celebrity status of being an astronaut's mother. His grandfather had also committed suicide years earlier. Like Nowak, Aldrin, too, had a loving but demanding parent—his father—who had high expectations for his son but withheld even "his smile of approval."

"I had flown to the moon and gotten its dust on my feet. But what was there for me now? What new goal could I set? What could possibly top that accomplishment?" Aldrin wrote. "I was physically exhausted and emotionally drained . . . I felt a growing sense of meaninglessness in my work as an astronaut, knowing that I would never return to the moon, or even fly another mission anywhere . . . How could I have gone almost overnight from being on top of the world to feeling useless, worthless, and washed up?"

He said that by the fall of 1970, he was frustrated and anxious.

"Surprisingly, however, nobody from NASA, and no medical or scientific study group, has ever analyzed the emotional aftereffects of space travel, especially the effects of instant celebrity and the pressures of a public life on those who were pilots or scientists," Aldrin wrote.

He left NASA and returned to the Air Force, hoping a stint as commandant of the United States Air Force Test Pilot School at Edwards Air Force Base would cure his depression. Instead, it worsened. Unbeknownst to the public, Aldrin secretly spent a month in late 1971 undergoing psychiatric treatment at Brooks Air Force Base's medical center in San Antonio, Texas, under the care of the chief of psychiatry, Colonel John Sparks.

Aldrin also admits to heavy drinking and an affair. He divorced his first wife, Joan, and retired from the Air Force. He told *Psychology Today* magazine in 2001 that he had to reexamine his life and its purpose.

"I had a genetic tendency toward alcoholism [both parents]. That eventually reached its peak in 1976 and 1977. Recovery was not easy. Perhaps the most challenging turnaround was accepting the need for assistance and help," Aldrin said. "Looking back at it now—with [decades] of sobriety—this was probably one of my greatest challenges. But it has also been one of the most satisfying because it has given me a sense of comfort with where I am now."

He has served as the chairman of the National Mental Health Association and is currently advocating for manned missions to Mars, along with space tourism. He also works with schools to educate the next generation of space pioneers.

In 2007, Dr. Jonathan Clark told FOX News that NASA does not do enough to monitor the mental health of its astronauts.

"They don't have to have any evaluation before or after a mission. It is only when something catastrophic happens that this ever comes to light," said Dr. Clark.

Dr. Clark also appeared on CNN, defending Nowak and saying she should be given the benefit of the doubt.

"Sometimes mental illness can be compounded by situations," Dr. Clark told Nancy Grace. "And she obviously had been through quite a bit after *Columbia*. She was one of our casualty assistance officers and helped immensely with my son's dealing with his loss. She then had to face the reality of going into space and facing that same risk that the other astronauts do. It's not unlike flying in combat. It's a very risky endeavor. And so she had to prepare herself and her family for that potential outcome. [You] come back from this immense high from spaceflight, and there's this void there that now happens after you've been through something that's really incredibly exhilarating, and then there's this void. And sometimes, you continue on in other thrill-seeking endeavors, such as, maybe, you know, the things that she was—has been accused of."

Homer Hickam, a retired NASA engineer whose life story was told in the book *Rocket Boys*, which was made into the movie *October Sky*, said NASA usually kept astronaut troubles quiet.

"This isn't the first case of astronauts having difficulties in their personal lives," Hickam wrote after Nowak's arrest. "Usually, the straying astronaut simply resigns or retires, and everything is hushed up. But being charged with assault, attempted kidnapping and attempted murder is far greater than anything I ever observed or imagined could occur. Perhaps this tragedy will bring some of the agency's long-ignored problems into the open."

Hickam pointed out that there were too many astronauts and not enough flights—something NASA also said in its 2007 Astronaut Health Care System Review. It made for a toxic, highly competitive work environment. In addition, ground crews composed of engineers and scientists often resented all the attention foisted upon the astronauts.

"Does it make sense to have this many overachievers all walking on eggs, vying for such a limited number of slots?" Hickam asked. "Only in a dysfunctional bureaucracy like NASA's astronaut office, which keeps hiring more astronauts than it needs. As a training manager, I was aware that many astronauts felt as if they were powerless, stressed-out peons within their own organization."

Dr. Santy, a vocal critic of NASA's behavioral health issues, is all too familiar with this pattern. On her blog in 2007, Santy wrote: "Why go to the bother of choosing 'the right stuff' in the first place when the superstar culture of the astronauts only encourages the worst sort of narcissism and sociopathy? Even if astronauts didn't have an iota of such psychopathology before their selection as an astronaut, they are at extremely high risk in the toxic NASA culture of developing it."

In a 2010 telephone interview, Dr. Santy said that NASA did not even keep the mental health evaluation records from the Mercury, Gemini, and Apollo programs—she found them in a file cabinet in someone's Houston garage.

"NASA itself required that the public image of astronauts be that of heroes and actively encouraged that public perception. Heroes many of them were, but definitely not superhuman," Dr. Santy said. "This perceived need to place the selected candidates unsullied before the public was an important reason why psychiatric, psychological and behavioral issues were not addressed in a straightforward or scientific manner within the space agency. NASA was so concerned with maintaining and protecting this image that all psychiatric data collected during selection were expunged from the official NASA medical records of astronauts."

According to Dr. Santy, the number of hours spent psychologically evaluating astronaut candidates went from thirty hours for the Mercury program in the 1950s, to ten hours for Gemini and Apollo candidates in the 1960s, to just *three hours* for the space shuttle program. The psychologists also went from military personnel to contractors hired by NASA—on their payroll, with files kept in NASA offices—everything under their complete control and review, with no outside influences and no outside prying eyes.

"NASA's reluctance to accept and deal with the importance of psychological issues—despite the accumulated evidence in every other hazardous environment and despite numerous efforts on the part of many scientists to enlighten them—is reminiscent of the denial and arrogance manifested by NASA managers prior to the launch of Challenger in 1986," Dr. Santy wrote.

As NASA transitioned to hiring space shuttle astronauts, Dr. Santy said there was little, if any, attempt to identify any specific psychopathology in any of the candidates. "The psychiatric consultants appeared to believe that their role should be that of determining who, in their judgement, was best suited to be an astronaut."

In addition, family members were not and are not interviewed to verify the veracity of candidate answers. People seeking top-secret government clearance see a far more rigorous process, with numerous family members, work colleagues, and neighbors spoken to by investigators.

The 2016 *Evidence Report* states that "important aspects of an individual's mental health history—e.g. exposure to a traumatic event, family history of mental health struggles such as depression or schizophrenia—are not always discoverable during the selection process. Not only may potential astronauts be hesitant to share information that would prohibit selection, but also, some current astronauts have demonstrated a reluctance to share information if they perceive such information could jeopardize their flight status, limiting the utility of countermeasures available to them."

The report adds that "despite careful selection, a depression-free past does not guarantee a depression-free future."

In the 1987 space shuttle astronaut applicant group, one with which Dr. Santy dealt, those for whom psychiatric disqualification was recommended included:

"An individual with both an anxiety disorder and a schizotypal
    personality disorder,
an individual with a narcissistic personality disorder,

and an individual with an adjustment disorder and a compulsive personality disorder."

In 1989, at least one person who met the criteria for having numerous hypomanic episodes was selected as an astronaut by NASA. "It is not known at this time if a history of hypomanic episodes is predictive of later psychopathology (e.g., bipolar disorder)," Dr. Santy wrote.

What's remarkable is that the Soviets—now the Russians—used one of NASA's early psychological screenings as the basis for their screenings, and they continue to do so. Dr. Santy recalled when Dr. George Ruff—the originator of NASA's psychological screenings—attended an international aerospace meeting in Prague in 1966 and met his Soviet counterpart, Colonel Kuznetsov.

"Kuznetsov refused to believe that the Americans were no longer doing in-depth psychological research on astronauts and the effects of space flight," Dr. Santy wrote. "He thought the Americans were 'pretending' not to be doing anything. Ruff had to convince the Soviet that, it was, in fact, true.

"'What a great shame,' Kuznetsov said."

However, Dr. Santy points out that, although the Russians spend a considerable amount of time screening cosmonauts and pairing those they feel will get along well, "the number of anecdotally reported interpersonal or emotional problems on Soviet missions has been rather high." Once, she said, the crew refused to talk to ground control.

NASA documents show that several times, Russian spaceflights were terminated early due to psychological factors, including a complaint of a pungent odor, the source of which was never found and no other crew smelled it. Soviet officials felt that because the crew had not been getting along, they experienced a "shared delusion or Folie à quatre."

The United States has not been without its issues.

The Skylab 4 mission in 1973–74 was eighty-four days long, with four spacewalks, but there was little time off built in and a lot of complicated tasks to perform, including some for which they had not trained long or well. So, according to the *Los Angeles Times,* Gerald Carr, Edward Gibson, and William Pogue turned off communication with Houston around December 28 and spent the day either looking out the window or on Skylab's solar console.

A 2016 *Evidence Report* states that there are three behavioral medical conditions included in the Lifetime Surveillance of Astronaut Health: behavioral emergency, depression, and anxiety. They are considering adding adjustment disorder.

According to an Associated Press article in 2007, NASA has a set of written procedures for dealing with an astronaut in space who develops psychological issues. Handling behavioral emergencies takes up five pages of the 2007, 1,051-page document, *Astronaut Health Care System Review.*

"The astronaut's crewmates should bind his/her wrists and ankles with duct tape, tie him/her down with a bungee cord and inject him/her with tranquilizers of necessary. Talk with the patient while you are restraining him . . . explain what you are doing and that you are using a restraint to ensure that he is safe."

There are no guns or stun guns in space, the AP reported. "NASA has determined there is no need for weapons on the space station," a NASA spokesman said in 2007.

A flight surgeon in Houston and the commander on board the shuttle would decide whether to end the flight or send an "unhinged astronaut" home if the episode took place on the ISS.

Space station medical kits currently contain tranquilizers and antidepression, anti-anxiety, and antipsychotic medications.

The NASA checklist says astronauts who have a mental breakdown can be restrained and offered oral Haldol, an antipsychotic drug used to treat agitation and mania, and Valium. The drugs can be forcibly given via a shot in the arm. Crew members are instructed to stay with the tied-up astronaut to monitor vital signs.

"Space station astronauts talk weekly via long-distance hook-up to a flight surgeon and every two weeks to a psychologist, so any psychiatric disorder would probably be detected before it became so serious that the astronaut had to be brought home," the Associated Press reported.

As far as has been reported, no NASA astronaut at the space station has ever been treated with antipsychotic or antidepressant medication, and no NASA shuttle crew member has required antipsychotic medication.

The AP reported that U.S. astronaut John Blaha said he was depressed at the start of a four-month stay on board the Soviet *Mir* space station in 1996.

"I think you have to battle yourself and tell yourself, 'Look, this is your new planet . . . and you need to enjoy this environment,'" Blaha told The AP. "You sort of shift yourself mentally."

According to the 2016 NASA *Evidence Report,* there has never been a mental breakdown in space, nor has anyone been diagnosed with depression or anxiety while on board a space shuttle or the ISS.

But out of 208 crew members who flew on eighty-nine shuttle missions between 1981 and 1989, there were thirty-four incidents of behavioral signs and symptoms reported—most commonly "anxiety and annoyance." During an unspecified time frame, flight surgeons and crew medical officers have documented twenty-four incidents of anxiety-related symptoms presented in spaceflight, and four astronauts experienced signs and symptoms of depression during spaceflight.

Major life events on Earth can negatively impact an astronaut's mental health. Just before Christmas 2007, Daniel Tani's ninety-year-old mother was killed in a car accident while he was on the ISS—and days after his mission had been extended by four months.

"Living on the space station means that I experience all aspects of life—be they joyous or tragic—while circling the Earth without a convenient way to return," Tani wrote in a press release at the time. He chose to continue with his duties as scheduled, although NASA offered him some time off.

In an Associated Press story that week, NASA spokeswoman Nicole Clouter said the space agency wasn't concerned about Tani's ability to perform his duties even after his having received such devastating news.

There have been incidents that gave astronauts pause, particularly with payload specialists, who go through training to perform a specific task on board a shuttle mission, usually a science experiment, but someone who is not trained to fly the spacecraft and did not go through the same psychological screenings as the other crew members.

Commander Brewster Shaw put a lock on the hatch on the side of the space shuttle because he didn't trust payload specialist Rodolfo Neri Vela, who flew in November 1985. Shaw said it was the first time he had flown with someone he did not know well.

"I didn't know what he was going to do on orbit," Shaw said. "So I remember I got this padlock, and when we got on orbit, I went down to the hatch on the side of the Orbiter, and I padlocked the hatch control so that you could not open the hatch. I mean, on the Orbiter on orbit you can go down there and you just flip this little thing and you crank that handle once," he said, showing with his hands how it was done, "the hatch opens and all the air goes out and everybody goes out with it, just like that. And I thought to myself, 'Jeez, I don't know this guy very well. He might flip out or something.' So I padlocked the hatch shut right after we got on orbit, and I didn't take the padlock off until we were in de-orbit prep. I don't know if I was supposed to do that or not, but that's a decision I made as being responsible for my crew and I just did it."

Shaw acknowledged, though, that Vela was a "great guy."

That same year, Taylor Wang had an issue. He was a scientist and payload specialist who had spent two years training to study fluid physics on board the space shuttle. According to the 2016 NASA *Evidence Report,* on the second day of his mission, his experiment failed.

"In his own words, he panicked," the study states. "When he asked mission control for time to repair his experiment and was denied due to schedule restraints, he threatened that he was 'Not going back' to Earth."

Apparently Wang felt like his failure was a reflection on the entire Chinese community. Dr. Santy said Wang became deeply depressed and the crew put a guard by the hatch. In the end, his crew members took on some of his tasks, which allowed time in the schedule for Wang to repair his experiment.

The late NASA pilot and commander Hank Hartsfield Jr. flew on three shuttle missions from 1982 to 1985. He recalled an issue with another payload specialist.

"We had one payload specialist that became obsessed with the hatch," he said, recalling the man's words. "'You mean, all I got to do is turn that handle and the hatch opens and all the air goes out?' It was kind of scary. Why did he keep asking about that? It turned out it was innocent, but at the time, you didn't know. We had some discussions, so we began to lock the hatch."

In June 1991, that feeling of mistrust had still not evaporated. Bryan O'Connor requested and used combination locks on his flight, saying the two payload specialists on the flight "were not career aviators and had not gone through the same training and experiences as astronauts," he told NASA.

"It's because we don't know you guys all that well," O'Connor said to Drew Gaffney and Millie Hughes-Fulford, even though he had trained with them for two years.

The most recent possible behavioral health issue came in the summer of 2019. The *New York Times* said NASA began investigating whether astronaut Anne McClain hacked into the bank account of her estranged spouse, Summer Worden. Worden said she became suspicious when McClain began asking her questions about her finances and what she could afford. Worden, a former Air Force intelligence officer, said she had contacted her bank to find out the locations of computers that had accessed her account, and one stood out: a computer registered to NASA during McClain's six-month mission aboard the ISS from December 2018 through June 2019.

The case against McClain fell apart when the United States Attorney's Office in Houston unsealed a federal indictment against Worden in February 2020, claiming Worden had made false statements to NASA OIG and to the Federal Trade Commission. If convicted, Worden faces up to five years in prison on each count and a possible $250,000 maximum fine.

On the ground, Dr. Santy said in a 2019 phone interview that astronauts need to undergo evaluations after any major life event, including issues with marriages, families, children, and deaths.

"On this kind of a professional level, when you're dealing with life and death, those are the kinds of issues that they need to deal with," Dr. Santy said. "They're not immune to the same kinds of things that plague the rest of us in life . . . [T]hose are the kinds of things that have impact, especially in an environment so unforgiving as space."

Instead, she said, astronauts are currently allowed to handle it in their own way. For some, she said, that means flying their T-38 training jets.

"That's a $50 million machine—that's not exactly what it's there for," Dr. Santy said. "One astronaut, Dave Walker, after his wife left him, got into a T-38 and almost hit a passenger jet" in 1990 outside of Washington, D.C. His jet came within one hundred feet of the airliner, and he was grounded for a time.

Dr. Santy also blames NASA for the breakup of a lot of marriages, saying they have created an atmosphere of tolerance for divorce, one that does not exist in the military.

"Families are sacrosanct in the military, but at NASA there's almost an encouragement to break up families," Dr. Santy said, adding that when she worked at NASA in the late 1980s through the 1990s, she saw many marriages end.

"The astronaut office might do a little bit more to discourage this kind of affair—they don't," Dr. Santy said, adding that many male astronauts married younger women who worked at NASA as secretaries, engineers, support personnel, and other female astronauts. "They became one of our patients . . . we no longer took care of the wife, we took care of the girlfriend. The military frowns on fornication. Anyone who is having an affair is encouraged to see a psychiatrist. And they're grounded. You're not going to be flying multimillion-dollar planes while they're [having an affair]."

Dr. Santy said one astronaut couple, assigned to the same flight, secretly married nine months before their 1992 flight. Jan Davis and Mark Lee became the first-ever married couple in space and, some say, possibly the first couple to have sex in space. NASA has never confirmed that and also established a rule after their flight—no married astronauts on the same flight. Davis and Lee have since divorced and remarried others.

Dr. Jonathan Clark, the widower of astronaut Laurel Clark, agreed. He told *USA Today* in February 2007 that NASA has turned a blind eye to both astronauts' mental troubles and their extramarital affairs.

"Now you see this is the consequence of not dealing with it—you have someone whose life is destroyed," he said. "Maybe they'll start dealing with it."

Dr. Santy added that better screenings for every person flying on board a spacecraft, including any upcoming tourists, is a must. A handful of tourists have already gone up.

"What astronaut is going to tell you they're feeling homicidal?" Dr. Santy asked. "They're very conscious that if they say the wrong thing, they could get grounded."

While Aldrin said he knew he needed help, he also knew seeking it could be career-ending. So, instead, he sought treatment for "neck and back pain" and was admitted to a private room on the second floor with other neck injury patients at the Brooks Air Force Station Medical Center. But he slipped up to the fourth floor for daily psychiatric counseling during his month-long hospitalization.

"[I knew] if I sought treatment for my mental and emotional traumas, a report would surely make its way into my official permanent records, thwarting any hopes I might have for future promotion," Aldrin wrote. "In the early 1970s, certainly in the military, to let it be known that you were seeking professional help for mental illness or alcoholism or drug addiction was a death knell to your career and certain to ostracize you socially."

His Apollo 11 crewmate Michael Collins also spoke about his concern about being grounded.

He told the PBS *NewsHour* in 2019 that wearing his bulky, white space suit in the Apollo Command Module simulator in Houston gave him claustrophobia.

"I was wedged in below one of the couches and very limited space—I couldn't really move," Collins recalled. "I was almost trapped . . . I never confessed that to anybody at that time. I was afraid I would be grounded."

The 2016 *Evidence Report* noted that being grounded has long been a fear for pilots and astronauts alike.

"One NASA Behavioral Health and Performance researcher has related that more than one astronaut has informed him that they respond to psychological tests in such a way as to confuse or mislead the researcher," the report states.

One of the worst moments for NASA came with the disintegration of the space shuttle *Columbia* in 2003. According to *Texas Monthly,* in the aftermath of the accident, "there had been real psychological trauma at the agency: According to one NASA source, there were 'several nervous breakdowns' and, among support personnel, one suicide. It was seen as such a serious problem that therapists were unleashed."

"We have this immense loss that we can never undo," Paul Hill, the deputy director of mission operations for the space shuttle, told the magazine. "There was collective and individual counseling, and we brought in various cultural coaches and psychologist types who spent time talking to our folks and boring into our culture."

It is not known if Nowak received any individual counseling in the days, weeks, or months after the *Columbia* accident, although she was working around the clock to care for Laurel Clark's son, Iain, in addition to her own family. According to NASA, Nowak's medical review team did not examine her records following the incident in Orlando.

Since the final space shuttle flight in July 2011, NASA is training astronauts for long-duration trips on board the International Space Station and, eventually, back to the moon and on to Mars. Astronauts would fly in cramped vehicles to the moon the size of the largest SUV, or to Mars in something the size of a luxury bus on a trip that would last three years, if not longer.

While "shuttle astronauts were expected to just tolerate any stressors that arose during their missions and were successful at doing so . . . 'ignoring it' is much less likely to be a successful coping mechanism . . . on longer missions."

According to the "Human Research Program Roadmap," a 2015 study, they have taken special care to address the psychological needs of the astronauts, dividing NASA's Behavioral Health and Performance Element into three areas:

1. the risk of performance decrements and adverse health outcomes resulting from sleep loss, circadian desynchronization, and work overload;
2. the risk of performance and behavioral health decrements due to inadequate cooperation, coordination, communication, and psychosocial adaptation within a team;
3. and the risk of adverse cognitive or behavioral conditions and psychiatric disorders.

According to the 2016 NASA *Evidence Report,* the agency "has a dedicated team of professionals whose primary goal is to provide psychological support to ISS astronauts and their families. The Johnson Space Center Behavioral Health & Performance group is a multidisciplinary team of mental health professionals. ISS astronauts are assigned their Behavioral Health and Performance team two years before their actual launch . . . Inflight services consist of: psychological support hardware, crew care packages, recreational materials, personal videos/photo uplinks, news and information uplinks, and private family conferences."

The 2016 *Evidence Report* also states that special care packages are sent up, with items provided by family, friends, the BHP team, and co-workers. Those packages can include "favorite foods, surprise gifts from the family, and holiday decorations . . . The most valued of these is the Private Family Conference. This deeply appreciated communication service occurs each weekend with videoconference equipment placed in the astronaut's home that allows private two-way video and audio between the astronaut and their family."

One of the most important supports for astronauts is the internet protocol telephone, which is actually a video conferencing system, with one end set up in the astronaut's family home and the other on board the ISS. It allows for private conversations.

ISS astronauts also have email capabilities, and "Crew Discretionary Events" allow astronauts "to make contact (audio or video) from the

ISS with extended family members, celebrities, military academies, and others that help boost the morale of the ISS crew."

However, in a long-duration mission, video uplinks will be difficult because of a twenty-minute delay in communication.

Astronauts' hobbies are also taken into account for those on long-term flights. A keyboard and guitar are permanent fixtures on board the International Space Station, with Canadian Commander Chris Hadfield famously playing David Bowie's "Space Oddity" in a video that went viral. Astronaut Scott Kelly spent a year in space in 2015–16 and took thousands of photographs, which he posted frequently to his Facebook page. They are now compiled in his book *Infinite Wonder*.

"Seeing the Earth from space, the overview effect, is a source of comfort to astronauts, who have reported it gives them a feeling of protectiveness or sense of responsibility," said James Picano, senior operational psychologist with NASA's Behavioral Health and Performance group, in a 2018 article in *Discover* magazine.

Astronaut Mike Lopez-Alegria emphasized the importance of seeing Earth in a 2010 interview with National Public Radio: "Looking out the window and seeing the Earth below, and seeing places you recognize and where you grew up and places you visited has a lot to do with keeping sane, so to speak."

Seeing Earth from a heavenly vantage point has, according to the 2016 *Evidence Report*, contributed to astronauts and cosmonauts having "transcendental, religious experiences or a sense of the unity of humankind while in space."

Apollo 14 astronaut Edgar Mitchell, the sixth person to walk on the moon, was so moved by this transcendental experience, of understanding that all living things are connected, he founded the Institute of Noetic Sciences "to explore the interplay between scientific knowledge and inner knowing."

The view of Earth is so important that researchers recommended to NASA in the 2016 *Evidence Report* that some kind of virtual reality window be set up in any long-distance exploration mission so astronauts can still keep an eye on home.

Some astronauts have asked to have a personal plant in space, while others enjoy tending to vegetable plants and small gardens on board the ISS. Some astronauts have said having a plant and being able to smell it and the soil it's in helps to reconnect them with Earth.

Exercise is a vital part of any astronaut's day. It helps fight off the blues but also keeps bone-density and muscle loss at bay, serious problems with long-term spaceflight.

According to a December 2017 NASA document, "assessment of crewmembers' behavioral health status is primarily done by interviews with a psychiatrist or psychologist. Preflight, there is a general review of any life events that may impact behavioral health, as well as a review of the crewmembers' support system, and a general mood assessment. Inflight on the International Space Station, there are regularly scheduled private psychological conferences, monitoring of mood and evaluation of work/rest schedules. Postflight, there are regularly scheduled interviews to assess the crewmembers' psychological re-adaptation to life on Earth." However, those only last for six months.

There is a laundry list of factors that can contribute to the risk of developing behavioral and psychiatric conditions, including:

Personality, including how it relates to adjustment
Resiliency or hardiness
Physiological changes that occur when adapting to microgravity
    and isolation
Emotional reactions, especially negative emotions
Radiation exposure, which is always a factor in any spaceflight
Habitability and environmental design of the spacecraft
Job design—performing meaningful work
Monotony and boredom
Daily hassles
Major life events missed on Earth
Cultural factors
Ground support/mission support
Family and social support
World events
Lighting and sleep shifting, with the resulting disruptions to cir-
    cadian rhythms.

Zero gravity, high carbon dioxide rates, and radiation exposure—all of which currently occur on the ISS and will occur on a long-duration mission to Mars—affect the human body in significant ways.

Zero gravity can contribute to motion sickness. Astronauts report vision issues—sometimes hallucinations of flashing lights, possibly caused by sustained motion sickness and insufficient blood flow to the brain.

Bone-density and muscle loss has long been a serious concern. Astronauts must exercise regularly to keep their bones as healthy as possible.

Carbon dioxide exists on the ISS at a rate ten times higher than levels on Earth. It could contribute to confusion and lethargy.

Radiation exposure is the greatest risk astronauts face, the 2016 *Evidence Report* states. As any cancer patient knows, radiation exposure "significantly degrades cognitive performance"—in NASA's case, to the point that the mission could be compromised. "Ionizing radiation damages the central nervous system, which leads to fatigue, negative mood, as well as difficulty sustaining attention." It can also speed up the effects of any neurodegenerative disease, like Alzheimer's.

"Deep space exploration will likely result in astronaut exposure to galactic cosmic radiation, consisting of high-energy, high-charged particles that are known to pose a significant threat to the brain and cognitive abilities."

Both radiation and stress can affect the brain's hippocampus, the area vital to the consolidation and retrieval of long-term memories. Radiation can change the hippocampus's neuronal structure, plasticity, and architecture. NASA currently uses the Spaceflight Cognitive Assessment Tool for Windows (WinScat) on board the ISS to test for cognitive processes and memory.

NASA is also researching the effects of diet on radiation exposure, including eating flaxseed, dried plums, blueberries, and strawberries to prevent or fight off radiation effects, the 2016 *Evidence Report* states.

Scott Kelly, who spent a year in space to study its effects on him, described what his life was like when he came home during testimony before Congress in June 2016, three months after his return.

"My muscles more quickly stiffened, and because my skin had not touched anything for nearly a year, it was extremely sensitive and became inflamed," Kelly told the House Subcommittee on Space. "I developed a hive-like rash on every surface of my skin that came in contact with ordinary surfaces on Earth during normal activities like sitting or lying in bed . . . I had flu-like symptoms that appear to have been a result from my extended time in space."

With a return to the moon and Mars missions looming on the not-so-distant horizon, NASA is concerned about the mental health of its astronauts. The 2016 *Evidence Report* states: "While mood and anxiety disturbances have occurred, no behavioral emergencies have been reported to date in space flight. Anecdotal and empirical evidence indicate that the likelihood of an adverse cognitive or behavioral condition or psychiatric disorder occurring greatly increases with the length of a mission. Further, while cognitive, behavioral, or psychiatric conditions might not immediately and directly threaten mission success, such conditions can, and do, adversely impact individual and crew health, welfare, and performance."

In fact, when comparing long-duration spaceflights with data regarding behavioral problems from Antarctic missions, NASA researchers stated in the 2016 *Evidence Report* that there is, at a minimum, a 53.4 percent chance that "a serious behavioral problem will occur" during the long-stay option in a crew of six over nine hundred days, with a maximum 89.3 percent chance. "Behavioral problems here are defined as symptoms that normally would warrant hospitalization."

Just as alarming—a NASA Office of Inspector General report released in October 2015 "noted that, as of August 2015, NASA does not have a validated mitigation strategy for any of the behavioral risks for a Mars mission."

Several factors contribute to astronauts' mental well-being while on board a spaceship: exercising, performing meaningful work, eating together, and having helpful crewmates.

A 2018 study, "Teamwork and Collaboration in Long-Duration Space Missions: Going to Extremes," in the journal *American Psychologists* shows that a sense of humor is an important component of today's "right stuff."

"Humor . . . is often cited as a benefit by spaceflight and analog teams, although sometimes it can cause friction," the study states. Astronauts say "that appropriate affiliative humor is a key factor in crew compatibility, conflict resolution, and coping—'humor and joking around continue to be huge assets and quickly defuse any problems.' Groups, particularly those in isolation, tend to develop their own cultures, complete with internal jokes, which bind them together."

Although, the study states, in some cases this results in conflict with other groups, such as between astronauts and Mission Control personnel.

Another factor for a positive contribution to a mission is having the trait of being "salutogenic." A Russian researcher defined it as having a sense of coherence in which "one has a pervasive, enduring though dynamic feeling of confidence that one's internal and external environments are predictable and that there is a high probability that things will work out as well as can reasonably be expected." Factors contributing to salutogenesis are understanding what is happening, managing their reactions, finding meaning in their life, having a social support, and engaging in spirituality, happiness, humor, and love.

As for taking tourists into space, Dr. Santy said in 2019 that officials should "say no. That's what they should've done for all of those high visibility" flights, like with Teacher in Space Christa McAuliffe and former U.S. Senator Bill Nelson, D-Florida.

"They feel like they have to let Congressmen fly," Santy said. "Anybody *cannot* do it."

She said U.S. Senator Jake Garn, R-Utah, was "basically sick the entire mission," when he went up in April 1985. Even though he was a U.S. Navy fighter pilot, he didn't have astronaut training. "He never got beyond that. Over time you can pick up skills to cope with this and he didn't—he was sick the whole time."

She likened it to allowing congressmen to perform surgery using a new procedure that Congress is thinking of funding.

"No—we wouldn't think of doing something like that," she said. "They're not only a danger to themselves, they're a danger to others when they're in an environment they're not able to cope with."

She said one of the things that upsets her with NASA is that "they have promoted the idea that it's all fun and games in space—it is a great adventure . . . [T]he teachers in space and the payload specialists really don't have a deep appreciation of the dangers in it. They don't know how much they're risking their lives . . . It's a very dangerous enterprise. It's not all blowing bubbles in zero gravity—bad things happen."

Another recommendation made by the 2016 *Evidence Report* was for retired astronauts to undergo a visit with a psychiatrist as part of their annual health screening at Johnson Space Center—60 percent of former astronauts participate in a lifelong health study. For some, a visit with a psychiatrist could be life-saving. There has been one attempted suicide by a French astronaut and at least one retired astronaut suicide since 1958.

On July 23, 2006—a week after Lisa Nowak and her crewmates finished their spaceflight—Dr. Chuck Brady died of self-inflicted knife wounds on Orcas Island in the woods near his home in Washington

State. He was fifty-four and had retired from the U.S. Navy four months earlier. He had been an Eagle Scout, a member of the Blue Angels flight demonstration team, a sports medicine doctor, and an astronaut who flew on one shuttle flight in 1996, when he conducted experiments on board space shuttle *Columbia*.

NASA conducted an investigation after Brady's suicide. His colleagues and his fiancé, Susan Oseth, were interviewed, and his annual medical evaluation from 2005 was reviewed. According to an online biography, Brady had suffered for years from severe arthritis and had undergone a hip replacement surgery.

"My interest in investigating this further was to look for lessons learned, in terms of identifying and acting on an astronaut with psychiatric problems," one person wrote in a November 2006 email. The names of who wrote the email and who received it were blacked out. "Did we miss something . . . should we have intervened and acted at some point?"

But, another email points out, "there are no BHP [behavioral health and performance] clinical records, but [redacted] personally reviewed the selection file and psychological testing."

Brady's selection file and psychological testing were from 1992— fourteen years before his death.

Brady's friend and colleague, biologist Loretta Hidalgo Whitesides, wrote eloquently about Brady in *Wired* magazine the next year, describing him as passionate about the environment and noting that she often talked with him about the connections between Earth and space.

Hidalgo Whitesides implored readers that, as we venture back to the moon and on to Mars, "Let's not forget that the human system can be just as complex as the mechanical ones. It also needs thought, care and attention. It is critical to the success of our mission. It may not be easy or comfortable to discuss or to work out, but then again, we go into space, 'not because it is easy, but because it is hard,'" she wrote, quoting President John F. Kennedy.

28

★　★

Possible Insanity Defense

On August 27, 2007, six and a half months after the attack in the airport parking lot, Nowak's attorney, Donald Lykkebak, filed a motion in the Orange County, Florida, Circuit Court to reserve the right to use an insanity defense.

"This notice does not challenge competence to stand trial, but only raises insanity at the time of the offense," Lykkebak wrote in the filing.

Two psychiatrists had evaluated Nowak in the months following the incident. Dr. Richard Pesikoff is a Baylor University School of Medicine doctor most famous for successfully convincing a jury that Andrea Yates was insane at the time when she drowned her five children in her Houston suburb home. Yates, suffering from severe postpartum depression, was convinced that Satan was inside her and she was saving her children. He saw Nowak twenty-nine times, starting February 24, 2007.

Dr. George Leventon's primary area of expertise is in child psychiatry. He also serves as an expert witness. He had two multihour visits with Nowak.

In the court filing, Lykkebak went on to name a laundry list of mental conditions from which Pesikoff and Leventon said Nowak was suffering at the time of the incident. They included:

Obsessive-compulsive disorder;
Partner relational problem;
Major depressive disorder (single episode, severe);
Insomnia related to obsessive-compulsive disorder;
Brief psychotic disorder with marked stressors (producing a
    mixed manic and depressive like state);
Asperger's Disorder;
Recent loss of 15 percent of body mass;
Problems with primary support group;

Marital separation;

Unable to confide in members of family of origin;

Problems related to the social environment;

Relationship problem;

Inadequate social support system;

And unable to confide in social contacts.

This list of diagnoses amounted to one thing: the doctors seemed to conclude that Lisa Nowak had temporarily lost her mind at the time of the attack.

"During the six weeks prior to the incident with Captain Shipman, Captain Nowak had been sleeping only 2–3 hours per night," Pesikoff wrote in a report to the court. "In addition, by the time her meeting with police began, she had been awake for almost 24 consecutive hours. Furthermore, she had recently lost 19 pounds and had almost nothing to eat during the 24 hours preceding the police questioning. The psychological and physical effects of fatigue, sleep deprivation and rapid weight loss, and poor nutrition . . . can lead to a variety of symptoms, including confusion, poor judgement, disorientation and psychosis."

Other experts reviewing the case, but who had not treated Nowak, disagreed that she had lost her mind.

Dr. Toni Werner was a psychiatrist with the University of Florida (UF). She had also served as an expert witness in many trials. Werner graduated from the American University of the Caribbean, then located in Montserrat. She did her residency in Connecticut and received a fellowship in forensic psychiatry from the University of Florida. Her conclusion: Nowak had some mental disorders, but she was not insane.

In Florida, in order to be declared insane, someone must not know that what they're doing is wrong, Werner explained. Clearly, she said, Nowak knew what she was doing was wrong because she tried to cover up her actions—checking into the DeFuniak Springs Days Inn under an assumed name, paying cash for gas, food, and lodging, wearing a disguise, and trying to hide the disguise from police.

"The problem with the insanity defense with her is she did so much planning beforehand and her actions were more based on jealousy," Werner said. "It's not really based on a delusion or psychotic thinking."

She compared the case to that of a murder in a town near Gainesville. Shortly after the September 11, 2001, terrorist attacks, a man was sitting with his mother. He later told police that his mother "smelled like a terrorist" so he killed her before she could kill him.

"He was delusional," she explained. "[Nowak's] actions were based on reality. Someone can be mentally ill and commit a crime—it doesn't mean they're insane."

One by one, she reviewed the findings by the Texas psychiatrists, referring to the definitions and characteristics of each one as defined in the *Diagnostic and Statistical Manual IV*. At the time, version IV was the bible for psychiatric evaluations, although it has since been updated.

She explained that diagnoses are made on five axes: (1) major mental psychosis; (2) personality disorders; (3) medical issues; (4) stressors; and (5) an assessment of functioning, ranging from 0 to 99, with 99 being the healthiest.

"No one is a 100," she said.

At the time of the event, according to the doctors who evaluated her, Nowak was at 30. By the time she was evaluated, she had improved to 65.

Listed under the major mental psychosis for Nowak: partner relational problem.

"It indicates that's the focus of therapy," she said. The definition includes "a pattern of interaction between partners or spouses with negative communication, distorted communication or non-communication."

"That's a lot of marriages," Werner added wryly.

Also listed: major depressive disorder.

"Most likely, in all honesty, she had a major depressive disorder with psychotic factors," Werner said.

The attributes of major depressive disorder include: a depressed mood more days than not, lack of interest, guilt, feelings of worthlessness, a diminished ability to concentrate, decreased energy level, disturbances with sleep (either too much or too little), suicidal ideation, and a significant weight loss when not dieting, with a change of more than 5 percent.

Nowak's weight loss of twenty pounds in a one- to two-month period was 15 percent of her starting weight of 127.

Psychotic features can include hearing voices, visual hallucinations, paranoia, and illness delusions—such as, "I know I have cancer." There are two other: erotomanic delusions—like John Hinckley shooting the president to get Jody Foster's attention . . . or driving more than nine hundred miles to "talk to" or—allegedly try to—kill your romantic rival so you and your boyfriend can be reunited. The other psychotic feature is jealousy.

Pesikoff concluded that Nowak had no intention of harming Shipman but that she became obsessed with the idea of helping Oefelein.

"Her mental condition rapidly deteriorated in January 2007 when she learned about 'the other woman,'" Pesikoff wrote. "After observing that Commander Oefelein was depressed and withdrawn, Captain Nowak began ruminating extensively; ultimately developing the delusional belief that it was her job to make Commander Oefelein happy. She felt she could accomplish this goal by having a meeting with Captain Shipman. Her thinking at the time demonstrated evidence of grandiosity, manic depressive features, sleeplessness and delusional thinking."

The next axis evaluated by psychiatrists is any kind of personality disorder. Nowak had two: obsessive-compulsive disorder (which the Texas doctors had actually listed under major mental psychosis) and possible Asperger's disorder.

Obsessive-compulsive disorder (OCD) has a number of symptoms, according to the book, many of which Nowak displayed.

People with OCD are preoccupied with details, rules, lists, order, and organization. Nowak obsessed over the details of her job, fine-tuning every aspect of it, to the exclusion of helping others with their jobs on board the space shuttle and International Space Station (ISS). And while most people would just throw items into a bag for a quick trip, Nowak's list of the items to pack in her duffel bag (gun, knife, pepper spray, jeans, sneakers, black gloves, etc.) is a classic display of OCD behavior.

OCD sufferers are perfectionists to the point that it interferes with completing a task. After years of training, Nowak perfected her use of the robotic arm on board the space shuttle and ISS. But she could not complete other tasks requested by crewmates, possibly out of a fear that she would not perform perfectly something she told them she had "not trained" to do.

Dr. Leventon noted that, at home, she was a "pack rat." Her house was filled with useless detritus that she refused to get rid of. When she tackled a hobby, she didn't just buy one or two items or spend just a few hours a week on things. She became obsessive. Her collection of rubber stamps numbered in the hundreds. When she and her husband went on a diet, she created a computer program to track daily vitamin and caloric intake from fat, carbohydrates, and protein. The program also factored in individual body mass and age. When her sister told her that a computer program for that already existed, Nowak called her back a few hours later, citing ten technical reasons her program was superior.

In addition, Leventon said, when he had to leave the consultation room to take an emergency phone call, Nowak began trimming "about a dozen leaves she felt were beginning to turn yellow off my African violets." She had also given him a gift of one of her 400 plants and cuttings.

People with OCD are extremely devoted to work to the exclusion of friends. In the months leading up to a launch, astronauts have very little time for anything other than work.

OCD sufferers are overly conscientious, unable to discard worn-out objects, have a miserly spending style, and are reluctant to delegate tasks to others.

They also have a rigid stubbornness. That could explain the incident at NextFest when Nowak refused to let a worker retrieve a pen for autographs from a desk drawer.

The third axis was any medical issues, like Nowak's loss of body mass. Biochemists explain that a major weight loss over a short period can throw off a person's chemistry, plummeting them into illness and even mental illness. Nowak wasn't eating. She wasn't sleeping. It was a recipe for a quick plunge into depression and a long drive across the country.

Nowak was also diagnosed with a form of autism—Asperger's disorder.

"Captain Nowak exhibited the social naiveness, the literal interpretation of facts, and the lack of appropriate social responses when dealing with the rejection she received from Commander Oefelein; all typical features of Asperger's Spectrum Disorder," Pesikoff wrote. "She was never consciously aware or capable of recognizing how hurt she felt by his rejection of her for another woman."

Dr. Michael Herkov is the director of forensic and addictive psychology at the University of Florida, specializing in Asperger's. He said he had been paying attention to the Nowak case because of the diagnosis, but had not treated her.

Herkov said it was possible for her to have a mild form of the disorder and still be an astronaut, still function, to a certain degree, in society.

He described Asperger's as a pervasive developmental disorder.

"It's not something you get; it's something you are," he said. And it usually begins in childhood.

"Their cognitive development and language is normal—that's one of the ways to differentiate them from autism," he said.

But what the two syndromes or disorders share is a difficulty or inability to socialize normally or read body language.

"If I were in an elevator and you got on, everybody would shift their position and you would pick a space somewhere in the middle," he explained. "A person with Asperger's would go right next to you."

And that would explain her odd behavior in nursery school, her coworkers' descriptions of Nowak as prickly, and her social interactions as odd.

Herkov said that, although Asperger's patients make social interaction blunders, it is possible for someone with a mild form to date and even marry.

"As they become adolescents, they tend to recognize these things and compensate," he said.

Leventon noted that Nowak's family told him that, as an adult at family gatherings, it was typical to find Nowak talking with her nieces and nephews or her own children, but she was uncomfortable interacting with the grown-ups in the room. They called her "clueless" when it came to social situations, but very competent in structured work settings. They said she rarely called to speak to them—they were always the ones who had to call Nowak. Unless she had a specific question for them. And she often did not return their phone calls.

Pesikoff noted that Nowak had poor eye contact and avoided social situations because she "felt shy and awkward." She watched very little television and only used the internet to shop so she could avoid being around people. He added that a number of people described her as a "very non-physical and non-violent person. She has never been in a fight, does not lose her temper, has no criminal record, does not use illegal drugs or drink alcohol, and does not spank her children."

But it's also hard for someone with Asperger's to disengage from a relationship—or rather for people involved with them to disengage. They don't pick up on the subtleties.

Oefelein met Colleen Shipman in November 2006, and they began dating. But he didn't tell Nowak for about seven weeks. Instead, he began limiting his contact with her. He couldn't deal with a breakup in November because he was preparing for his shuttle flight in early December. After he landed, on December 22, Nowak called him nearly two dozen times in the first few days—calls he did not answer or return. And then Christmas came.

Nowak told Pesikoff that "Oefelein told her he still loved her but needed time to think about their relationship."

Finally, in the beginning of January, he sat her down and told her that he had met someone and had fallen in love.

He was doing a classic "fade out," but people with Asperger's don't understand that, Herkov said.

According to the two Texas psychiatrists, Nowak also had a number of stressors, or additional factors, that led to her mental breakdown.

She had, in many ways, isolated herself—or been isolated—from her primary support group—her family, her closest friends, her colleagues, her neighbors, her husband. Although they all loved Nowak and thought she was a wonderful person, she did not confide in any of them that she was having an affair with Billy Oefelein. Nor did she discuss her breakup with him—how could she since she was also breaking up with her husband at the same time? And many of them also didn't know about that.

Dr. Leventon noted that she had a difficult time connecting to women, citing her intense privacy as a possible reason why. Even her friendship with Laurel Clark, who died in the *Columbia* accident, was not what it could have been.

"Even with Laurel, she was reticent to divulge her inner thoughts," the psychiatrist wrote.

Like the military version of *Mean Girls*, the wives' club of her military squadron invited her to a meeting but then blackballed her—she wasn't the wife of a squadron member . . . she was *in* the squadron.

Her parents were angry about her separation and pending divorce. In a letter to Billy Oefelein's mother that is undated, but clearly written in the weeks before the attack—and before she found out about Colleen Shipman—Nowak wrote about Mrs. Oefelein's kindness, noting that her parents were not as supportive of her. Shipman said a similar letter was eventually mailed—even after a court order instructing Nowak to have no contact with Shipman or Oefelein.

People with these issues, UF's Werner explains, can have delusions, feelings of guilt, self-punishment or persecution, nihilistic behavior, or somatic responses.

"It depends on their underlying coping mechanisms," Werner said.

"She didn't have a structure at home, she didn't have a support partner or even her mom. She was turning to this gentleman and when that started to go down, she lost everything. She had nothing left."

And, Werner pointed out, Nowak was failing in multiple areas: her marriage had failed, she had failed to secure a second shuttle mission—essentially ending her career as an astronaut—and she had failed in her relationship with Oefelein. Not only was failure not an option for her, it was something she had rarely experienced.

At what point did she think it was a good idea to drive from Houston to Orlando, carrying a BB pistol, knife, mallet, pepper spray, plastic gloves, surgical tubing, trash bags, and a disguise?

"By the time she came up with the idea, she was having the erotic delusions," Werner said. "She didn't have any checks and balances—she didn't have anybody to confide in who might say, 'That's not a good thing.' Because of the mental illness, she was unable to see it was not a good plan. Although she knew what she was doing was wrong."

So, was the woman who launched into space on July 4, 2006, the same woman who drove across the country seven months later, who police and prosecutors said wanted to kill a romantic rival?

"She was the same woman—with a mental illness," Werner said.

However, there is no reason why she could not go on to live a normal life, with the proper treatment and, possibly, medication.

"People who have chronic mental illness, like schizophrenia, you see a downward drift," she said. But with Nowak—"there's no reason why we wouldn't expect her to regain her [pre-incident] level of functioning."

Nowak's depression, like that of thousands of others in Houston, probably began in February 2003 during one of the worst moments for NASA—the disintegration of the space shuttle *Columbia*.

Despite the extensive diagnoses, Nowak's attorney eventually withdrew his intent to use the insanity defense.

29

★ ★

Legal Maneuvers

Colleen Shipman immediately asked for and was granted a restraining order against Nowak, saying she had been stalked for two months prior to the incident and she feared for her safety.

Within a month of the incident, the state attorney's office filed three formal charges against Nowak: (1) attempted kidnapping with intent to inflict bodily harm or terrorize; (2) burglary of a conveyance with a weapon; and (3) battery.

There was some good news for Nowak—prosecutors did not file the attempted murder charge that had been recommended by Assistant State Attorney Darlene Simmons. Nowak's attorney entered a written plea of not guilty to the charges; Nowak did not attend the hearing. Judge Marc Lubet scheduled a trial to begin in July.

Experts all over the country were weighing in on the case. Former Los Angeles Police detective Mark Fuhrman, who rose to fame as an investigator on the O. J. Simpson case, said the items in the duffel bag Nowak carried with her were actually a "murder kit." Those items included a BB pistol, hunting knife, 2½-pound steel mallet, surgical tubing, three pairs of latex gloves, large garbage bags, and a disguise.

"If you have any one of those objects that she had, even a wig, they are innocent all by themselves," Fuhrman told *Florida Today*. "When you buy all these items, and the packaging is in the back seat of the car . . . it collectively paints a much more incriminating picture. We have a suspect . . . going on a shopping spree to go murder somebody."

In April 2007, the state attorney's office released more documents in the case, including police reports indicating they found a floppy disc in Nowak's car that contained fifteen photographs and drawings of women in various states of bondage. In addition, there were hand-drawn maps to Shipman's house, schedules, and a checklist of items to bring on her trip, including: BB pistol, a knife, disguise, plastic gloves, a hat with hair, and black gloves.

Also found in her car: several fortunes from fortune cookies. One read: "The coming month shall bring winds of change in your life."

In April, there were some behind-the-scenes maneuvering going on. Orlando Police detective Chris Becton had reached out to Wade Krieger, an investigator with NASA's Office of the Inspector General (OIG), for help on cleaning up the recording of his questioning of Nowak.

"Would it be possible for NASA to attempt and clean up the audio from the Nowak interview?" Becton asked in an email. "Her volume is way lower than the background noise. Our Audio/Visual people could not improve the quality of the audio."

NASA contractor Indyne was going to help with the recording, especially in spots where Nowak had answered, but the recording was garbled.

But Lykkebak learned of this development and contacted NASA's chief counsel at Johnson Space Center, Bernard Roan, to object to it.

"If true, I find this extraordinarily unusual and inappropriate," Lykkebak wrote in an email. "The people NASA may be assisting wish to put her in prison for the rest of her life. Perhaps NASA engineers would like to drive the paddy wagon to the state penitentiary? It seems improper for the employer to assist in the employee's execution."

Roan wrote to several people at NASA Headquarters in Washington, D.C., that Johnson Space Center's inspector general "confirms we're going to do it—because [defense] counsel complained he couldn't hear what was on the tapes. Moreover, we (thru requests to the IG . . .) routinely do it."

But within ten days, the offer of help from Indyne was off the table.

"Given the high level briefings required, as you suggest, I believe it's best that we let OPD pursue a commercial vendor," NASA special agent in charge John Corbett, based in Houston, wrote in an April 20 email to a higher-up at NASA headquarters in Washington, D.C.

Krieger broke the news to Becton via email: "As we discussed on the phone, per the e-mail below, although we do have an account with Indyne, which was set up specifically to fund this kind of work on open NASA OIG cases, HQ will not allow me to use our account to fund this request. What can I say? It's political. Sorry man—you guys will have to find a commercial vendor on your own to get the work done."

In May 2007, Lykkebak asked Judge Lubet to postpone the trial as he awaited a clean copy and transcript of the surveillance video of Nowak in the airport holding cell. When asked by reporters what he thought was on the video, Lykkebak bristled.

"What makes you think I would tell you?" *Florida Today* reported him saying. "Leave me alone, will you? I'll try my case in the courtroom and not in the hallway."

At the end of June 2007, Lykkebak filed two motions to have Nowak's statement to police and evidence found in her car declared inadmissible. He said Becton did not correctly read Nowak her Miranda rights to remain silent and to have an attorney. In addition, he said Becton bullied her into telling him where her car was.

"As such, these admissions constitute fruit of the poisoned tree," Lykkebak said.

Outside the courtroom, Lykkebak launched into a tirade against the media and claimed Nowak never told Becton she used diapers while on the trip.

"The biggest lie is this preposterous tale," Lykkebak growled to reporters, including me. "A lie repeated over and over and over again can overcome the truth."

He did admit that used diapers were found in her car by Becton. In a court filing, he said they were toddler size 3, cartoon character diapers. But, he said, the diapers were used in September 2005 by the Nowak family, including two toddlers, when the family had to evacuate Houston for Hurricane Rita. He said the family wound up at a hotel that was full and the management would not allow nonguests to use the hotel's facilities. The Nowak family stayed in their car in a crowded parking lot and used the diapers in the car rather than urinate in front of others in the parking lot. He did not explain why, if this were the case, she would have left soiled diapers in the family's car for a year and a half.

Becton stood by what he wrote in his report and maintains, to this day, that Nowak told him she used the diapers during her cross-country trip so she would not have to stop. Becton theorized that she also did not want to be seen in public.

Lykkebak said he also would be filing a motion to have Nowak's ankle monitor removed. Shipman's attorney, Kepler Funk, said he would object to that.

"Ms. Shipman still has fear of Miss Nowak," Funk said. "If you ask people who have been victims of personal violence, that fear takes sometimes years—and sometimes never—to go away. Consequently,

I will object to any loosening of the reins that the court has on Miss Nowak."

In August 2007, Lykkebak made good on his promise to file a motion on the removal of the monitor. Nowak was paying $105 per week for the device and called it an "unnecessary and excessive expense." She added that it interfered with her ability to exercise, which is a requirement for a Navy officer, and her ability to drive, fly on a commercial plane, and monitor her children in the pool.

"The device is not small, comfortable, lightweight or unobtrusive, as its supplier would have one believe," the filing stated. In addition, Nowak said it was a safety hazard because the low-battery alarm had gone off on her commute to Corpus Christi on a rural two-lane road. She is required to change the battery within two minutes or be in violation of her bail requirements.

On August 24, 2007, Nowak presented those same arguments in court as she testified. It was the first time that Nowak and Shipman had been in close proximity since the attack.

But Shipman said she wanted the monitor to remain on Nowak's ankle.

"When I'm home alone and there's nobody there with me, it is a comfort," Shipman told Judge Lubet. She was wearing the gold space shuttle charm that Oefelein had given her as a gift.

But she also said that she had gone to Houston to see Oefelein four or five times since the February attack.

"You left the comfort of that GPS monitor, which you thought would protect you in Brevard County, to travel exactly to the state and city where you knew she lived?" Lykkebak asked, incredulous.

Lykkebak spent much of the hearing trying to pick apart testimony from Detective Becton. He argued that Becton improperly read Nowak her Miranda rights to remain silent and to have an attorney, and that he denied her access to an attorney. Lykkebak also said Nowak never gave consent to search her car.

"It was the hardest interview I've ever conducted in my career," Becton testified. "I realized I was dealing with somebody who was more intelligent than I was—more educated . . . I was looking at the interview as a chess match."

The transcript of his interview shows that he did read her the Miranda rights—just not all at once. He also told her several times she had the right to an attorney and that he would have to stop questioning her if she asked for one. She repeatedly asked him if he thought she needed an attorney, but she never outright asked for one. Becton would answer by pointing out how smart she was.

As for the car search, Becton testified that, after five hours of questioning, Nowak gave consent by saying, "um-hummm" and nodding. She also gave a slip of paper with the location of the blue BMW, guiding him and an officer to the La Quinta Inn near the airport. And then she gave him the keys.

Becton also testified about the most talked-about aspect of the case—the diapers. He was startled when he found them, thinking she had brought her children with her.

"She said she used the diapers in order to pee so she wouldn't have to make so many stops," Becton testified under oath.

"My impression was the car would be searched, no matter what," Nowak later said on the stand, also under oath.

Outside the courtroom, more than a dozen reporters gathered around a lectern and waited for Nowak to appear to make a statement.

"The past six months have been very difficult for me, my family and others close to me," she said. "I know that it must also have been very hard for Colleen Shipman, and I would like her to know how very sorry I am about having frightened her [in] any way, and about the subsequent public harassment that has besieged all of us."

Nowak also talked about the overwhelming attention paid to her and Shipman by reporters—both mainstream and tabloid.

"I've been both shocked and overwhelmed at the media coverage," she said, describing a press "invasion of my street" and describing reporters ringing the doorbells of friends, neighbors, and coworkers at all hours, and parking their news and satellite trucks on people's lawns.

But she also said she was grateful to the people who have supported her through the ordeal, including family members who had given her "unconditional love."

"One thing that has surprised me throughout these months has been the number of people whose generosity, insight and support and motivation has come shining through this cloudy time. The hundreds of people who wrote me personal letters," she said, getting a little choked up. "I've read and kept every single one . . . It would have been very easy for me to permanently retreat into a world of personal sorrow, but my family and friends have given me a greater view about what is important in our lives, so even though life may change suddenly and drastically, there can be a lot of great yet to be accomplished. I don't know yet how to do that, or if the final outcome of this case will allow it, but I do know that I have these amazing people to thank for making any future possible."

Shipman's lawyer, Kepler Funk, said the statement "sounds like an admission of guilt."

A week later, Judge Lubet ruled that Nowak could remove the ankle monitor.

"It's a great relief not to worry about safety issues related to the battery's life while I'm driving," Nowak said through her attorney. "I'm also really looking forward to getting back into my former aerobic fitness programs."

Funk was outraged.

"In our opinion, justice requires that Lisa Nowak's aerobic fitness should not override Colleen Shipman's peace of mind," Funk said.

Around the same time, the U.S. Navy ordered Nowak to stay away from Shipman and Oefelein.

Chief of Naval Air Training D. B. Grimland said "violations of this order may result in administrative or disciplinary action under the Uniform Code of Military Justice." Grimland added that the step was taken to "ensure the safety and protection" of Shipman and Oefelein. "It is also intended to protect you from further allegations while this order is in effect."

In September 2007 another hearing was held, this time to determine if Nowak's statement and possessions found in her car would be allowed to be entered into evidence during her trial, which had been postponed until this issue was settled.

Becton testified under oath that he thought he would find evidence in Nowak's car to further support his theory that "Ms. Nowak actually came here to kill Colleen Shipman."

He said the weapons found in the duffel bag were "items that have been used—and can be used—to kill somebody" and that he thought he would find tools to "bury Shipman in a shallow grave."

At issue was also the way in which Becton read Nowak her Miranda rights to remain silent and have an attorney. She insisted that she did ask for a legal counsel.

"My memory, that I recall, is that I did say that, but it's not in the transcript," Nowak testified.

Lykkebak said that when Nowak asked several times whether or not she needed an attorney, all questioning should have stopped.

She also swore under oath that she did not give Becton permission to search her car. But she did give him the location of the vehicle and the keys to it.

"I absolutely did not ever say, 'You could search the car,' and I did not give my consent," she said.

Prosecutor Pam Davis said Nowak's portrayal of herself as "basically a puddle of indecision and confusion" was false and described the aeronautical engineer as a rocket scientist who was smart enough to determine if she needed a lawyer.

"She doesn't have to be a legal officer to understand, 'I have the right to remain silent,'" Davis said. "That's basic English . . . She's got to say, 'I want a lawyer.'"

Lykkebak accused Becton of "cajoling and manipulation" by subjecting her to "prolonged detention," questioning her for more than four hours, threatening to withhold help if she didn't cooperate, and making promises to ensure mental health counseling.

Davis pointed out that law enforcement officers have a right to lie to suspects in order to gain information.

Judge Lubet eventually ruled that both her statement and evidence found in the car were not admissible in court.

"It is incumbent upon this court that every citizen, whether a prince or a pauper, be treated equally," Lubet wrote. "In each and every case, the court must ensure that the Constitutional protections afforded by our forefathers are scrupulously honored. Unfortunately, in this case, those protections were not as thoroughly followed as the law demands."

Lubet said Becton failed to answer Nowak's repeated questions in a "simple and straightforward manner" concerning her need for a lawyer, that he minimized Nowak's Miranda rights by referring to them as "formalities," and that Becton didn't utilize a written waiver-of-Miranda-rights form or a consent-to-search form. He also said Becton essentially bullied Nowak by requesting she speak to him before reading her Miranda rights to her, making false promises of help if she talked with him, threatening her with retaliation if she didn't talk, and refusing to allow her to make a phone call.

"Although [Nowak] is intelligent, this was the first time she had been arrested," Lubet wrote in his decision.

"I knew after the suppression hearing I was going to lose the interview," Becton later said. "It was my fault for not using the conference microphone. But my case didn't hinge on the chess match. The evidence in the car was the icing on the cake for me."

But he said that he was stunned when he read Judge Lubet's findings that Nowak didn't understand her Miranda rights.

"A rocket scientist, a fighter pilot, did not understand if she wanted an attorney?" he asked.

And then he read that Judge Lubet threw out the car evidence, too.

"I called Assistant State Attorney Pam Davis, and as soon as she heard my voice, she already knew what I was going to say," Becton recalled. "She assured me she was appealing the decision."

On January 16, 2008, Judge Lubet postponed the trial and the pretrial hearing scheduled for March indefinitely, pending the outcome of the state's appeal on his decision to throw out her statement and evidence found in her car.

On October 21, 2008, Davis appeared before the Fifth District Court of Appeal in Daytona to argue the appeal.

"I think it's clear she knew she had a right to an attorney," Assistant Attorney General Kellie Nielan, arguing for the state, told the five-judge panel. She pointed out that Nowak repeatedly asked Becton if she needed an attorney as he informed her of her right to have one.

The judges questioned both Donald Lykkebak and Nielan about the Carroll Doctrine, a legal principle that allows police officers to search a car without a warrant if there is "reasonable suspicion" of criminal activity. Becton had previously testified he thought he might find grave-digging equipment in the car.

"Given what they had recovered from her at the scene and statements from Ms. Shipman, was that really a stretch?" Judge Richard Orfinger asked Lykkebak.

Six weeks later, the five-judge panel held that Nowak's statements were taken in violation of her Miranda rights but that the search of her car was still admissible under the "inevitable discovery exception" to the search warrant requirement. They said the police would have eventually found her car in the normal course of the investigation.

On April 1, 2009, the judge ordered Nowak to undergo two psychiatric evaluations before June 12, 2009. She had visits with psychiatrist Dr. George Leventon, whose primary area of expertise is in child psychiatry, and Dr. Richard Pesikoff, a Baylor University School of Medicine doctor.

In May, Lykkebak dropped his use of the insanity defense.

On October 7, 2009, Lubet ruled in favor of allowing Nowak's attorneys to take a second deposition from Shipman to inquire whether Nowak actually pepper-sprayed Shipman. A medical report by paramedics raised some questions, according to Nowak's attorneys, as to the factual basis for it. If it was found not to have occurred, Nowak's attorneys wanted the criminal charges related to the assault and battery to be dropped before trial began. The trial was scheduled for December 7, 2009.

Finally, on November 10, 2009—two years and nine months after the incident—Nowak arrived at court to enter a guilty plea as part of an agreement with the state attorney's office. As reporters mingled in the hallway, waiting for the courtroom doors to open, Nowak, dressed in a black suit, blue blouse, and black pumps, arrived with Lykkebak; his friend Chaney Mason, an attorney who had helped on the case; and the team's crisis communications expert, Marti MacKenzie.

"Are you glad this is almost over?" I asked.

"Don't ask her any questions," MacKenzie ordered.

"It's my job to ask her questions," I said, smiling at MacKenzie.

"It's my job to tell you not to," she said, smiling back.

MacKenzie and I had held numerous off-the-record discussions during the time between the arrest and Nowak's final criminal court appearance. During many of those conversations, she firmly and repeatedly asked that I stop including the diaper issue. I didn't comply, but I always added that Nowak's attorney denied that part of the incident happened.

Nowak only stared straight ahead as everyone waited for the courtroom doors to be unlocked.

The former astronaut entered a guilty plea to the lesser charges of burglary of a conveyance, a third-degree felony, and a misdemeanor battery charge, as part of the agreement with the district attorney.

Shipman was allowed to speak and implored the judge to give Nowak the maximum sentence.

"Shortly after I turned 30 years old, Lisa Nowak hunted me down and attacked me in a dark parking lot. Her attack was part of a well-researched, well-planned and deliberate crime," Shipman told the judge as she read from a statement, her hands shaking as she occasionally teared up. "I know in my heart, when Lisa Nowak attacked me that she was going to kill me. It was in her eyes—a blood-chilling expression of limitless rage and glee. It's my understanding that Lisa Nowak had researched 'murder,' 'corpse dismemberment,' as well as disguises and trace evidence and I am 100 percent certain that Lisa Nowak came here to murder me and I believe that she never thought that she'd get caught. It was only by the grace of God and quick footwork that I escaped."

Shipman described high blood pressure and dizzy spells, chest pains and uncontrollable shaking, migraines, and nightmares.

"I lost my job and my military career," she said, adding that she had resigned from the Air Force and moved to Alaska with Oefelein.

"Lisa Nowak's claim that she only wanted to talk to me is, at best, ridiculous. I believe it is one of many lies that she designed to deceive and gain sympathy from this court, NASA, the U.S. Navy, her friends, her family, and the American people," Shipman said. "Please don't be fooled. I was fooled. Lisa Nowak is a very good actress. On February 5, 2007, Lisa Nowak had ample opportunity to talk to me. She stealthily followed me inside the Orlando International Airport for hours. I'm a very friendly person, your honor, who, at that time, would've welcomed her company over some hot chocolate while I was waiting for my suitcase to arrive."

As for the attack itself, Shipman said she was lucky to still be alive.

"When Lisa Nowak began following me through the parking lot, I noticed that she still had her duffel bag with her," Shipman said. "She was close enough and it was quiet enough that I could hear the swishing of her pants and I never heard her drop that bag. I had to sprint to my car to escape her. Had I been but a fraction of a second slower in locking my car, this court, I have no doubt, would instead be hearing a first-degree murder trial . . . I believe I escaped a horrible death that night."

Shipman described suffering from post-traumatic stress disorder from what she perceived as an attempt on her life.

"I have barricaded my doors, put in an alarm system, purchased a shotgun, and got my license to carry a concealed weapon. All in an effort to feel secure again," she said. "None of this has worked."

She described Nowak's flagrant violation of a judge and military commander's orders to have no contact with her or Oefelein. She told Judge Lubet that, in the fall of 2007, she found a book that Nowak had sent to Oefelein that contained several purple Post-it notes with sexually suggestive messages.

Shipman returned to her seat between her attorneys, Kepler Funk and Keith Szachacz. Her mother sat nearby.

In the weeks prior to the hearing, several of Nowak's friends had written to Judge Lubet on her behalf.

Her space shuttle crewmate Piers Sellers described her as an intelligent and hardworking crew member, who was kind, gentle, and considerate. He also called her a devoted and capable mother "who spent as much time as she could with her children."

"Lisa was a very conscientious and skilled robotics officer/flight engineer who, quite literally, held my life in her hands," Sellers wrote. "I was never concerned at any time that I was anything but completely safe in those hands, and I still feel the same way today."

Fellow astronaut Catherine Coleman said she and Nowak often exchanged childcare tips when their children were little and helped each other with picking up their children from school or daycare. Their children continued to play with each other. She said the incident was completely out of character for Nowak.

"I trusted Lisa with my son before this incident and I continue to trust her with his welfare," Coleman wrote. "Our interactions have not changed, and my respect for her and the way she cares for her children—and my son—has not changed either."

As Nowak and Lykkebak stood at the podium, Judge Lubet asked if Nowak had anything to say.

"I think she does, Judge," Lykkebak told the court.

A reserved Nowak spoke: "I'm glad to have this opportunity to apologize to Ms. Shipman in person."

Judge Lubet stopped her.

"Why don't you turn and face Ms. Shipman when you do this?" Judge Lubet asked her.

Nowak turned to face Colleen Shipman, speaking directly to her for the first time since February 5, 2007.

"I'm glad to have the opportunity to apologize to you, Ms. Shipman, in person. I am sincerely sorry for causing fear and misunderstanding and all of the intense public exposure that you have suffered," Nowak said, nodding her head. "I hope very much that we can all move forward from this with privacy and peace."

Lubet warned Nowak the judgment might affect her Naval career and military retirement benefits, which she could lose for pleading guilty to the third-degree felony.

"But you brought this on yourself, and I don't have any sympathy for you in that respect," Judge Lubet said before withholding adjudication, which means Nowak was not convicted of the crimes, although she was sentenced for them and it was placed on her record. He sentenced her to two days in jail, which she served when she was arrested. He also placed her on one year of probation and told her to do fifty hours of community service, go through anger management classes, and write "a sincere letter of apology to Shipman." In addition, she was ordered not to travel to the three states where Shipman and Oefelein live, have family, or work, nor can she ever have any contact with them.

In the elevator on the way out of the building, Shipman declined to comment, but Funk said she was pleased. "She got what she wanted and that was for Mrs. Nowak to admit what she did."

30

★ ★

JAG

On August 19, 2010, Nowak was back in Florida, this time at Jacksonville Naval Air Station on the banks of the St. Johns River. She arrived just before 7:00 A.M., wearing a pink sweater set, navy pants, and pearls—not her U.S. Navy officer's uniform.

A panel of three admirals gathered to review evidence, statements, and depositions in the case and to hear testimony from Nowak, her supervisor, and a doctor.

Rear Admiral Mark Boensel, Rear Admiral Eleanor Valentin, and Rear Admiral Timothy Matthews sat opposite Nowak in the hearing room. Captain Karen Fischer-Anderson and Lieutenant Peter Ostrom served as Nowak's counsel and assistant counsel.

The hearing was not a court-martial but rather an inquiry to determine whether Nowak should be separated from the U.S. Navy for misconduct, including the commission of a military or civilian offense, and substandard performance of duty, including a failure to conform to prescribed standards of military deportment.

William McDonald, a psychiatrist who had evaluated Nowak, was sworn in as a witness. The hearing room was cleared of all spectators to protect Nowak's privacy. When he concluded, everyone was brought back.

And then it was Nowak's turn to tell her side of the story. She began by detailing a long and storied career in the Navy, including flying more than thirty different kinds of aircraft and logging more than 3,100 hours of flight time during flight school, and serving as mission commander of electronic warfare school and electronic warfare lead, aggressor squadron VAQ-134, Naval Post-Graduate School, and Naval Test Pilot School.

Then she discussed her time with NASA and her "Return to Flight" mission aboard space shuttle *Discovery*. She talked about enjoying visiting schools and talking to children and how she did not like speaking with the media, but she endured it because she knew it was a part of the job.

"At one time, the children's father had asked me about the attention that we get as astronauts and that being excessive, compared to, for instance, the other people that work at NASA," Nowak said. "And certainly, that's my opinion, too. Everyone that works there is a team [member] and everybody's important."

And then she addressed the most controversial part of the incident and police report: the diapers.

"I've never worn diapers except for our duties for spaceflight when we were required to wear them during the launching sequences," Nowak told the panel. "The idea that I was wearing astronaut diapers or any diapers on that trip was false."

She went on to explain that she had a box of unused diapers in her car, along with used diapers in a garbage bag, dating back to a hurricane evacuation in 2005 when she and her family were forced to stay in their car in a hotel parking lot and use the diapers for urination because the hotel would not allow the family inside. Nowak had also told one of her psychiatrists the same thing, not expecting his report to ever be seen publicly.

She said she had described to Detective Becton how, "using two hands to hold this small toddler diaper in the area to collect [urine], that is exactly how it was used, and I described it to him. He later took that conversation and made it appear in his arrest report as if I had done that during this weekend, which is completely not true."

She also told psychiatrist Dr. George Leventon in August 2007, and the Navy panel in August 2010, that she thought Oefelein was upset about being in a relationship with both her and Shipman simultaneously. And so she wanted to help him solve this dilemma.

"He was gaining a lot of weight, and he appeared to be distressed, and I thought he was drinking more alcohol than would be normal for him," Nowak told the panel. "I wouldn't say he had stopped confiding in me, but we were still friends that would be able to trust each other with things like that."

She told them that she drove to Orlando in order to talk to Shipman about "his deteriorating health condition."

"I believed it would be beneficial to talk to her to help figure out what was wrong or what it was Commander Oefelein couldn't talk about with me," Nowak said. "I believed that he must talk with Ms. Shipman about some things, and I believed at the time that talking to her might provide an avenue to help lead to a road of recovery for him."

Not once in more than six hours of questioning or talking with the Orlando Police Department did she ever mention Oefelein being sick, upset, or depressed, or her wanting to help him.

She also said that, at the time of the incident, she feared Shipman and thought Shipman might hurt her if Shipman knew who she was. And that was why Nowak had carried some of the items in the duffel bag—she said she felt like she needed to protect herself from Shipman. But she also explained that she was not thinking clearly.

"I felt threatened in that situation because I had pre-developed this sense of fear, unreasonably and inappropriately," Nowak said when talking about pepper-spraying Shipman. "There was no reason, excuse or justification for spraying her window. I am very, very sorry and I accept all responsibility that comes with that."

However, in another part of her testimony, Nowak claimed that she had bought the knife and a sleeping bag for hunting trips with her son's Boy Scout troop and the BB pistol for her son to use in the Boy Scouts. However, a receipt shows she bought the BB pistol and knife together just a week and a half before her cross-country trip. And she never discussed Boy Scouts or fearing Shipman with the Orlando Police.

She said the surgical tubing was actually exercise bands, used in April 2006 after she sprained her ankle. And the steel mallet was bought to hang pictures because her husband had removed a lot of the tools from their home when he moved out in the weeks before the event.

Nowak said that, as Shipman drove away from her in the airport parking lot that night, Shipman almost hit her with the car. But she said she thought it was understandable, given the circumstances.

"I thought, 'What did I just do? Why did I do that? I don't know,'" Nowak recalled. And then all she could think of was: "I have to go home. I'm later than I wanted to be here. I need to go home now."

But she denied trying to get into Shipman's car—the most serious crime for which she pleaded guilty.

She explained to the panel that she decided to plead guilty to the third-degree felony because she wanted to spare her family further intense media scrutiny. She described her neighborhood after the event, with news live trucks parked up and down the street and on people's lawns, reporters harassing her neighbors and family at all hours, microphones shoved into her face at court hearings.

"It was going to be a long process of exposure for my family. I was worried about their safety, and weighing those factors against what now looked like the conclusion could be, there was no good resolution or answer to this," Nowak said. "There was no great way out, so what I had to do was weigh what would be the most good for the most people, to include my family, my children, my friends at NASA, the Navy and everywhere, with the least amount of harm to all those same people. Given all that, the media had a huge impact on making that decision."

Finally, she would not admit to having a romantic relationship with Oefelein. At the time of the incident, Oefelein told police it began in 2004 during a training mission. Both were married at the time, which made the relationship against military law.

"As far as to whether he and I had a romantic relationship, you can have that impression or not," Nowak told the panel. "If he's saying it was a romantic relationship that lasted two or three years, I'm not saying anything about that. I believed an interest expressed, and as far as following up on it, that was not considered possible under the circumstances, or appropriate, and in fact that is why he met and had developed a close relationship with Ms. Shipman, or at least that's what he told me."

James Hooper, chief of staff at Naval Air Training in Corpus Christi, then testified on Nowak's behalf. He had worked as an aerospace experimental psychologist for the Navy at one point. He described her work as excellent in standardizing thousands of training manuals and also streamlining a syllabus and flight training used for NASA personnel, reducing the number of days needed to train from ninety to just under thirty and making the work available remotely.

"Based on my observations, principally my direct observations of her contributions, her service has been quite satisfactory and most honorable," Hooper said.

According to *Florida Today*, Nowak was questioned about the items she had with her in the duffel bag. Navy prosecutors presented the black wig, BB gun, four-inch knife, rubber tubing, a steel mallet, and the pepper spray as evidence.

"During questioning by the government counsel, Nowak clarified again that she only had the spray and the wig with her but had left the other items elsewhere," *Florida Today* reported. "She said she had the other items because she feared Shipman might become violent."

After deliberations that same day, the panel voted unanimously to recommend Nowak be separated from the U.S. Navy with an "other than honorable" discharge and that her rank be reduced from captain to commander. A Pentagon official still needed to rule on the matter.

In March 2011, Nowak petitioned the civilian court to seal the record of her criminal proceedings, citing harm to her family and their livelihood. The motion was granted.

More than eleven months after the hearing in Jacksonville, Assistant Secretary of the Navy Juan M. Garcia III said in a statement that Nowak would retire with an "other than honorable" discharge and her pay grade would be bumped down one rank to commander.

"Captain Nowak's conduct fell well short of that expected of senior officers in our Navy and demonstrated a complete disregard for the well-being of a fellow service member," Garcia said in a written statement.

Although it could affect her ability to receive veteran's benefits, she was allowed to keep her retirement pay, estimated at about six thousand dollars per month.

Neither Shipman nor Oefelein were at the hearing in Jacksonville; they were on their honeymoon in Bora Bora.

31

★ ★

Postmortem

Former NASA psychiatrist and flight surgeon Dr. Patricia Santy, who did not treat Nowak and was not involved in her selection at NASA, said Nowak was very lucky with both the civilian case and the Navy hearing.

"I think she did get away with attempted murder and she's lucky she didn't succeed," Santy said. "If they're saying she had diminished capacity, that would be ridiculous—she planned things very carefully. She's lucky she didn't get court-martialed. In the past she might have been court-martialed."

And she placed some of the blame for the incident at the feet of Oefelein because of the affair.

"He's equally guilty," Dr. Santy said. "They both would have been equally culpable—he had an affair, too. He should've gotten the same kind of treatment. If he had been still married, he would've gotten the same treatment."

Dr. Santy said NASA and its culture of astronaut worship is also to blame.

"Of course you can't prevent everything, but something like the Lisa Nowak affair could've been prevented, very easily I think. Lisa Nowak had had problems in her marriage long before" the incident, she said in a telephone interview. "Those are the kinds of life events that should've been known to the people that monitor her at NASA—the flight medicine office. And that's the kind of secrecy that astronauts deliberately keep from them. They know deep down that kind of thing that affects them and keeps them grounded."

She reiterated that the unspoken rule at NASA has always been to tolerate almost anything the astronauts want to do. And, she said, at the very least, Nowak and Oefelein's crewmates knew what they were up to, but either ignored it or kept it quiet.

"I do think this is one of those red-flag psycho-social things that NASA should've had on their radar," Dr. Santy said. "If they had less tolerance for the freewheeling sex that goes on there, that would be a good thing. We used to hear the rumors . . . on this kind of a professional level, when you're dealing with life and death, those are the kinds of issues that they need to deal with."

From a legal standpoint, those reviewing the case for this book also felt as if Nowak got away with a far more serious crime than the third-degree felony burglary of a conveyance and misdemeanor assault to which she pleaded guilty.

Judge J. Michael Hunter, a retired circuit court judge from Polk County, Florida, looked through court documents and the transcript of Detective Becton's questioning of Nowak and also agreed that she "got off very light, considering how serious the nature of this was and what appears to be the nature of her intent."

Hunter said, based on the evidence in the duffel bag and some of her statements, that she clearly wanted to do more than talk to Shipman. Hunter quoted the appeals court ruling when the judges threw out her statement to Detective Becton but kept in the search of her car. They stated, in part, there was "evidence of a plan that likely extended beyond the airport parking lot."

"This court took the assumption that she was going to do more than she did," Hunter said. "I find it interesting that they added that part about evidence of a plan that likely extended beyond the airport . . . [T]hey thought that she planned to kill her without saying it. I guess it was just more than a law enforcement officer and a prosecutor that had that assumption . . . But I still say in terms of evidence to get there, you have to assume certain things that we don't have proof of. I believe that was probably what she was going to do, but I certainly couldn't prove it beyond a reasonable doubt."

While Hunter applauded the decisions of Judge Lubet, Nowak's attorney Donald Lykkebak, and the Fifth District Court of Appeals in Daytona, he saved any criticism for Orlando Police detective Chris Becton.

"You can tell he's an intelligent guy, but I sure think he handled it poorly," Hunter said of Becton's questioning.

From 1972 to 1982, Hunter worked in in law enforcement, serving as a sheriff's deputy in Lake and Polk Counties, Florida, and then as an agent with the Florida Department of Law Enforcement, the statewide police agency. He obtained his bachelor's degree in criminal justice from the University of Central Florida and a master's degree in it from Rollins College before teaching the subject at Northern Kentucky University for five years while attending law school at Chase College of Law at Northern Kentucky University in Cincinnati. Then he served as an attorney for the Polk County's Sheriff's Office before becoming a prosecutor and finally a judge, eventually overseeing death penalty cases.

"I'm more critical of police actions—he just didn't do it the right way," Hunter said.

"I'm most flabbergasted by the fact that he didn't read her the Miranda warnings using a preprinted form that all agencies like the Orlando Police Department have. He winged it," Hunter said. "Pull that thing out, read it to her out loud, and have her acknowledge she understands these rights and have her sign it . . . Do it right. I'm shocked that a veteran detective from an agency as allegedly professional as Orlando Police is did this. I'm equally flabbergasted that he started talking to her and carrying on a conversation with her before saying 'I need to go through some formalities.' He makes light of these constitutional rights by saying, 'I need to go through these formalities.' [And] he ran the risk of her giving him an admission before he read her her rights."

Hunter said the six-hour interrogation, after she had been kept by herself in a holding cell for two hours, when she hadn't slept or eaten, was also troublesome.

"To me, that's just inherently coercive, and then at that time of the day or morning to interrogate her for [six] hours, just adds to the coerciveness of the whole thing," Hunter said. "I'm troubled by the fact that he pressed her so many times—'Where's the car? Where's the car? Where's the car?'"

Hunter acknowledged, though, that Nowak was very wily in the interview, basically keeping Becton, whom Hunter acknowledges was an experienced detective skilled at interrogations, on the defensive.

"She did a very good job of talking for hours without saying anything," Hunter said. "She sure did a good job in spite of his—he gave her every reason why she should tell him what's going on and she didn't do it. I thought he was very convincing and he didn't convince her to do anything."

But, he said, her statement, in which she admitted only to pepperspraying Shipman, really didn't add a whole lot to her prosecution. He said police had her disguise, evidence she followed the victim, and evidence she was in the parking lot.

"She acknowledges she was there . . . I think there's ample evidence to put her at the scene of this crime and to prove she did what she did," Hunter said, "but none of her admissions were against her. She kept saying I needed to talk to her . . . that all sounds like someone with a little bit of a depraved mind at that time. But she never admitted—I think she said, 'I didn't mean to do her any harm,' so I don't see where any of her admissions hurt her."

A well-respected and longtime Florida prosecutor who asked not to be named came to the same conclusion, independent of Judge Hunter. He said the police statement wasn't needed and he would not have even admitted it into evidence had he been on the case.

"Statements are generally placed into evidence when they include culpatory comments—meaning they prove guilt," he said. "This statement of hers is exculpatory. She fought him every inch of the way—she was very tenacious. It's not what I would call a 'good statement.' She didn't admit guilt anywhere."

The prosecutor added that there was more than enough evidence to try to convict her, including a duffel bag filled with weapons, following Shipman in the parking lot, and then pepper-spraying her.

"I think the evidence is just as strong without the statement," he said. "A jury can give that the weight it deserves."

Judge Hunter was also surprised that Nowak did not utter the one phrase that would have stopped Becton in his tracks.

"She's an astronaut from Houston, Texas, and here she is in Orlando, Florida, put in a room for two hours to sit by herself and gets this skilled interrogator to come in and question her for [six] hours," Hunter said. "As an intelligent woman, I can't believe she didn't say, 'I need an attorney.'"

He applauded Donald Lykkebak's skill in convincing Judge Lubet to suppress her statement and the evidence in the car, saying he would have ruled the same way as Lubet because, though they would have eventually found her car, there was no probable cause compelling them to search it—other than Becton's hunch.

Judge Lubet "did an extremely thorough job of going through the case law on both of those issues and then he did a very thorough and diligent job of analyzing the facts as it related to the case law and then he came to, I think, the proper conclusion," Hunter said.

Lykkebak, he said, was an expert litigator.

"Thanks to her attorney, she cut a real sweetheart deal and to be able to keep her pension with the Navy," Hunter said. "I think she may have gotten the deal of the century."

Hunter agreed with Lykkebak in that Nowak's past record and history were important and that her stellar reputation did—and should have—helped her.

"Clearly she had a lot of 'attaboys' in that column—she was basically a hero," Hunter said. Before the incident, Nowak was "certainly a model for any woman regardless of any occupation they want, to go for it even if you're in a man's world."

32

★ ★

Where Are They Now?

Oefelein and Shipman both retired from the military and moved to his home state of Alaska. They created a website, Adventure Write, through which they have catalogued their explorations all over the world. They also sponsor an annual writing contest for students and work with the Forget Me Not Mission Foundation in Anchorage, Alaska, battling drunk drivers.

In 2009, in a remote spot in the wilderness on a 110-mile canoe trip, Oefelein proposed.

"He gets on his knee and says, 'Would you do me the honor of spending the rest of your life with me?'" Shipman told ABC News as part of an hour-long special. "I was floored—I was taken aback. And then he pulls out this giant ring."

In August 2010, Shipman and Oefelein married on a perfect summer day in a Pennsylvania garden. She wore a goddess-style chiffon gown with a plunging, rhinestone-and-pearl-embellished neckline. Her bridesmaids wore sage-green goddess gowns. Billy-O wore a black flight suit, as did the groomsmen and ring bearer.

At their wedding reception, the newly minted Mrs. Oefelein danced an Irish jig with her friends, donning the thick black shoes of her Celtic homeland and leaping at least two feet off the dance floor.

In November 2011, they welcomed their son, who frequently flies with them throughout Alaska, manning the controls for the first time in August 2019—under his father's supervision.

Oefelein has worked in the aerospace industry as a consultant. Colleen Shipman Oefelein published her first novel, a young adult paranormal book called *Eerie,* under the name C. M. McCoy. It received good reviews, has been prominently displayed in Alaskan Barnes & Noble stores, and is featured on the company's website.

Colleen Shipman Oefelein has done two interviews with *People* magazine and another, extensive interview with ABC News, tearing up at the thought of what might have happened that dark night in February 2007.

Billy Oefelein has never publicly discussed the incident or his relationship with Lisa Nowak.

Nowak's shuttle crewmate Piers Sellers flew again on space shuttle *Atlantis* in May 2010. He retired as an astronaut in 2011 and went on to become the deputy director of sciences and exploration at NASA's Goddard Space Flight Center in Greenbelt, Maryland. In 2011, he was appointed an Officer of the Order of the British Empire in New Year Honours for his services to science.

In January 2016, he wrote an opinion editorial on climate change for the *New York Times.*

"I'm very grateful for the experiences I've had on this planet," Sellers wrote. "As an astronaut I spacewalked 220 miles above the Earth. Floating alongside the International Space Station, I watched hurricanes cartwheel across oceans, the Amazon snake its way to the sea through a brilliant green carpet of forest, and gigantic nighttime thunderstorms flash and flare for hundreds of miles along the Equator. From this God's-eye-view, I saw how fragile and infinitely precious the Earth is. I'm hopeful for its future."

In that op-ed piece, he also announced that he had stage 4 pancreatic cancer. Piers Sellers died on December 23, 2016.

The pilot on Nowak's crew, Mark Kelly, married U.S. Representative Gabrielle Giffords several months after Nowak's incident. Four years later, Gifford was shot in the head during an assassination attempt when she was holding a "Congress on Your Corner" meeting at a shopping center. Six people were killed, and eighteen others were also shot by the paranoid schizophrenic gunman. Kelly and Giffords work to stand up for solutions to prevent gun violence and protect responsible gun ownership. In February 2019, Kelly announced his bid for the U.S. Senate in the 2020 elections.

Spacewalker Mike Fossum flew on two other spaceflights and served two missions onboard the International Space Station. He retired from NASA in January 2017 and was named vice president and chief operating officer of Texas A&M University's Galveston Campus.

Mission specialist Thomas Reiter flew one more mission on board the International Space Station. From April 2011 to December 2015 he was director of the European Space Agency's directorate of human spaceflight and operations. He is now the ESA's interagency coordinator and advisor to the director general.

Stephanie Wilson flew twice more to space—her forty-two days in low Earth orbit are the most flown by any African American astronaut. She remains a NASA astronaut.

Shuttle commander Steve Lindsey flew the final space shuttle flight in March 2011. He retired from NASA and lives with his wife in Lafayette, Colorado.

Lisa Nowak and her husband, Richard, divorced in June 2008. He gave her full custody of their three children. He continues to rent a home in Houston.

Lisa Nowak retired from the U.S. Navy on September 1, 2011, with the reduced rank of commander. She collects her retirement pay from the military and remains in the home she had once shared with Richard in Houston, raising her daughters and staying out of the public spotlight. Her son graduated from college and works as a web developer. A movie starring Natalie Portman, loosely based on the incident, was released in October 2019 to mixed reviews.

In 2007, Baylor University psychiatrist Dr. George Leventon wrote that "Captain Nowak presents no danger to herself or others . . . It is expected that, with treatment, she will achieve a complete recovery."

Marti McKenzie, a public relations specialist who worked with Nowak's defense team, declined an interview on Nowak's behalf, saying she will never publicly discuss the incident. McKenzie also declined interviews with Nowak's family members.

In 2012, Nowak attended a reunion of women astronauts; she was one of two dozen present. In a photo of the event, she is seen grinning the warm smile that mirrors her NASA portrait, taken before her shuttle flight and her drive into infamy.

Acknowledgments

This book is the work, in fits and starts, of more than a dozen years of interviews with dozens of people—both as a reporter and an author—along with substantial research and writing. I wanted this book to be a serious approach to help people understand how someone so accomplished and brilliant could descend into the grip of mental illness—and then recover to continue living a normal life.

My first thanks goes to former *Florida Today* assistant metro editor Don Walker for assigning me to cover the criminal case of astronaut Lisa Nowak, for always believing in me and my work, and encouraging me with corny jokes every day.

Orlando Police Department detective Chris Becton provided deep insight into this case and encouragement that sparked the idea to write this book. His and his wife's friendship is appreciated deeply.

The Harbaugh Salon Writers' Group—Pam Harbaugh and her late husband, John, Chris Kridler, Annette Clifford, Cathy Mathias, and Billy Cox—was invaluable in teasing out details, guiding me to an appropriate narrative, and encouraging me to keep writing, despite family issues and the theft of my laptop and external hard drive. Former *Florida Today* space reporter Chris Kridler was particularly helpful in sharing her notes from her preflight interview with Nowak and her crewmates, tweaking technical details about space shuttle launches, flights, and landings, pointing me toward more information, and being an enthusiastic supporter of this work from day one.

Thanks go to my good friend and fellow WSVN alumna Barbara Bendall for repeatedly dropping everything to do background searches on people and events when I couldn't find the information myself.

University Press of Florida has been on board with this project from its earliest days, when Amy Gorelick first saw a rough draft and suggested some wise changes. Editors Linda Bathgate and Sian Hunter then guided me through the acquisition and editing process. Susan Murray did the heavy lifting of copyediting every word in the manuscript, and I am grateful for all the changes she made. Thanks to Marthe Walters for herding my proofs edits.

Drs. Patricia Santy and Toni Werner, both psychiatrists, untangled the web of psychiatric diagnosis contained in Nowak's evaluations to help the average reader understand what she was thinking and doing before and during the attack on Colleen Shipman. Dr. Michael Herkov eloquently explained Asperger's disorder to me.

Appreciation goes to Jessica Cordero at NASA's Johnson Space Center for the always professional handling of my multiple Freedom of Information Act requests.

My brother, Mark Moore, generously provided a soft place to land as I finished the final phases of this work and moved on to the next chapter of my life. I will forever be indebted to him.

Several people reached out to me anonymously as I covered the case to help fill in gaps of information about Nowak and Shipman, and I thank them.

Retired Assistant State Attorney and U.S. Navy Judge Advocate General Commander Darlene Simmons provided a look behind the scenes and pointed to valuable information about the case that has been unknown publicly.

Retired Circuit Court Judge J. Michael Hunter gave judicial insight into the case and the legal proceedings that only someone who has dealt with death penalty cases can.

Many thanks to Colleen Shipman Oefelein for her help and encouragement, although she has her own story to tell.

Finally, I want to thank the tens of thousands of NASA employees, and those who work for its contractors, who carry out the important tasks of America's mission in space. Thank you for your service.

Bibliography

Primary Sources

Interviews by the Author

Becton, Chris. Orlando Police detective. Multiple in-person, telephone, and email interviews. 2007 through 2019.

Halvorson, Todd. *Florida Today* space reporter. 2008.

Herkov, Dr. Michael. Director of forensic and addictive psychology at the University of Florida. Telephone interview. 2010.

Hunter, Michael. Retired circuit court judge. September 2019.

Santy, Patricia, M.D. Former NASA psychiatrist. Telephone interviews. June 2010 and August 2019.

Werner, Dr. Toni. University of Florida psychologist. Interview in her home. Summer 2008.

NASA Materials

Advanced Crew Escape Suit (ACES). 2009.

Astronaut Code of Professional Responsibility. 2007.

"Astronaut Health Care System Review Committee Report FAQs." July 27, 2007.

Astronaut Healthcare System Review. June 2007.

"Behind the Scenes, Astronaut Candidates 2004 Training Journals, Eileen Collins." 2005.

Biography of Astronaut Anderson, Michael.

Biography of Astronaut Brown, Dr. David.

Biography of Astronaut Chawla, Kalpana.

Biography of Astronaut Clark, Dr. Laurel.

Biography of Astronaut Husband, Rick.

Biography of Astronaut Nowak, Lisa. March 2007.

Biography of Astronaut Oefelein, William. June 2007.

Biography of Astronaut Rqamon, Ilan.

Brady, Charles. Suicide investigation. 2006.

"Contingency Shuttle Crew Support Capability." April 28, 2006.

Emails of Lisa Nowak and William Oefelein.

Evidence Report: Risk of Adverse Cognitive or Behavioral Conditions and Psychiatric Disorders. April 11, 2016.

Evidence tag of previous lock-and-key set from William Oefelein's apartment. February 2007.

"How to Become an Astronaut.".

"Human Research Program Roadmap." 2015.

"Human Spaceflight, Living in Space." April 7, 2002.

"Ice Team." October 27, 2006.

Investigative interview of Dana Lois Davis, publicist for United Space Alliance.

Investigative interview of Astronaut Michael Fossum.

Investigative interview of Trent Herbert, flight crew operations support manager, United Space Alliance.

Investigative interview of Astronaut Mark Kelly.

Investigative interview of Astronaut Mark Polansky.

Investigative interview of Erlinda Lee Stevenson, scheduler for United Space Alliance.

Investigative interview of Renee Jacqueline Walker, publicist for United Space Alliance.

Johnson Space Center Astronaut and Flight Surgeon Survey Report. January 2008.

"Landing the Space Shuttle Orbiter at KSC." March 1992.

Living in Space. 2005.

Lyndon B. Johnson Space Center. *About Johnson Space Center,* 2000.

Medical Examination Requirements for Former Astronauts. December 12, 2017.

Mission Preparation and Pre-Launch Operations. 1988.

NEEMO 12 Mission Journal, May 2007.

Note of bit-for-bit copy of files of four astronaut quarantine quarters computers. February 2007.

Note of seizure of two astronaut office computers, IBM Thinkpads. February 2007.

Note of seizure of William Oefelein's home computer. February 2007.

Opening Remarks of Deputy Administrator Shana Dale Regarding the Astronaut Health Care System Review. July 27, 2007.

"Part Office, Part Sanctuary, All Crew Quarters." May 31, 2007.

Photo, STS-121 crew at The White House.

Preflight Interview with Mike Fossum. February 21, 2006.

Preflight Interview with Mark Kelly. February 23, 2006.

Preflight Interview with Steve Lindsey. February 23, 2006.

Preflight Interview with Lisa Nowak. February 21, 2006.

Preflight Interview with Bill Oefelein. November 3, 2006.

Preflight Interview with Thomas Reiter. February 23, 2006.

Preflight Interview with Piers Sellers. February 23, 2006.

Preflight Interview with Stephanie Wilson. February 23, 2006.

"The Real Survivors." May 20, 2004.

Receipt of evidence of USS Nimitz paper pad and United Airlines Mileage Plus Statement from William Oefelein's apartment.

Report of Columbia Accident Investigation Board. August 26, 2003.

Report to the President by the Presidential Commission on the Space Shuttle Challenger Accident. June 6, 1986.

"Robotics and Automation.".

"Sonny Carter Training Facility: The Neutral Buoyancy Laboratory." 2006.

The Space Medicine Exploration Medical Condition List. 2011.

"Space Shuttle Canadarm Robotic Arm Marks 25 Years in Space." November 9, 2006.

Space Shuttle Launch Countdown. 1995.
Space Shuttle Launch Sequence. 2000.
Space Shuttle Orbiter Systems. 1988.
Statement Regarding the Status of Lisa Nowak. March 7, 2007.
STS-116, "Preparing for Flight." November 17, 2006.
STS-121 Launch Blog, July 4, 2006.
STS-121 Mission Status Reports. July 2006.
STS-121 Press Kit. December 2006.
STS-121 Press Kit. "The Second Step." July 2006.
STS-121 Status Report. July 4–17, 2006.
"Suit Yourself Is Easier Said Than Done." December 22, 2008.
"This Week at NASA." September 2006.
Virtual Launch Control Center. July 2006.

NASA VIDEOS

Flight Deck video of *Columbia* prior to breakup. February 2003.
Flight Deck video of STS-121 Launch. July 2006.
Spacewalks of STS-121. July 2006.
SSP Continuing Improvements and Crew Profiles.
STS-121 Discovery Mission Handout.
STS 121 Feature Packages, Mission Overview, EVA Package.
STS-121 Post-flight Wrap-up.
STS-121 White Room video.
STS-121 Wrap-up video. www.youtube.com/watch?v=CWnDcbKKn34. 2006.

ORLANDO POLICE DEPARTMENT DOCUMENTS

Case narrative.
Charging affidavits.
Copies of written documents, including Shipman's flight itinerary, list of items to take
    on the trip, emails between Oefelein and Shipman, and a letter to Billye Oefelein.
DVD containing evidence photos and warrants.
Orlando Police holding cell video. February 5, 2007.
Phone records of William Oefelein and Lisa Nowak.
Photographs of evidence, including BB pistol, steel mallet, hunting knife, BB cartridg-
    es, knife packaging, wig, trench coat, hats, glasses, receipts, pills, cash, makeup,
    floppy disc, and flash drives.
Shipman-Oefelein emails.
Statement of Bus Driver Jean Marie Adrian.
Statement of Officer Timothy Ryan.
Statement of Parking Lot Attendant Maria Smith.
Statement of WESH photographer Frank Burt.
Supplement report.
Transcript of questioning of Lisa Nowak by Detective Chris Becton.
Transcript of questioning of William Oefelein by Detective Chris Becton.
Transcript of William Oefelein statement.

Video of Nowak following Shipman in airport.

Video of Nowak in holding cell.

Written statement of Colleen Shipman.

Court Documents

Appeal. November 2007.

Appeal decision. Fifth District Court of Appeal. December 5, 2008.

Charge sheet. February 6, 2007.

Defendant's Plea. February 13, 2007.

Deposition. Paramedic William Hagedorn.

Deposition. Paramedic Gregory Loebl.

Deposition. Colleen Shipman.

Indication of attorney. February 6, 2007.

Judge Marc Lubet's Ruling. November 10, 2009.

Letter to Judge Marc Lubet from Astronaut Catherine Coleman.

Letter to Judge Marc Lubet from Astronaut Robert Curbeam.

Letter to Judge Marc Lubet from Astronaut Piers Sellers.

Motion for possible use of insanity defense. August 27, 2007.

Motion in Limine with supporting memorandum. October 2009.

Motion to continue. September 1, 2007.

Motion to impose sanctions. October 9, 2009.

Motion to suppress admissions illegally obtained. June 29, 2007.

Motion to suppress evidence from unlawful search. June 29, 2007.

Notice of Appearance for Nowak. February 13, 2007.

Notice of Appearance for Shipman. February 13, 2007.

Notice of Discovery. February 13, 2007.

Order for reimbursement of investigative costs.

Order granting motion to suppress admission illegally obtained and motion to suppress evidence from unlawful search. November 2, 2007.

Petition for injunction for protection against repeat violence. February 5, 2007.

Other Primary Sources

Divorce Records. Harris County, Texas.

Human Spaceflight Ethics and Obligations: Options for Monitoring, Diagnosing and Treating Former Astronauts. Before the Subcommittee on Space, Committee on Science, Space and Technology. 114th U.S Congress.

Kridler, Chris. Former *Florida Today* space reporter. Preflight interviews with Lisa Nowak, including audio and transcripts, 2006.

Leventon, George, M.D. Psychological evaluation of Lisa Marie Nowak. August 19, 2007.

Marriage records. Anne Arundel County, Maryland.

NASA's Astronaut Health Care Systems—Results of an Independent Review. Before the Subcommittee on Space and Aeronautics, Committee on Science and Technology. U.S. House of Representatives, 110th Cong., 1st Sess. September 6, 2007.

Pesikoff, Richard, M.D. Psychiatric evaluation report of Lisa Marie Nowak. August 21, 2007.

Pronk, John. WFAA cameraman. Interview by WFAA. February 1, 2003.

Report to the President by the Presidential Commission on the Space Shuttle Challenger Accident. June 6, 1986.

Testimonies of Dr. Richard Williams, Retired Astronaut Scott Kelly, Chief Astronaut and U.S. Navy Capt. Chris Cassidy, Association of Space Explorers President Michael Lopez-Alegria. June 15, 2016.

Testimony of NASA administrator Michael Griffin. September 6, 2007. U.S. Congress.

Transcript of Lisa Nowak's U.S. Navy Board of Inquiry hearing to determine her future with the military. August 19, 2010.

U.S. Congress. News conference. Findings of Astronaut Health Reviews. July 27, 2007.

"U.S. Naval Academy Guide to Catholic Wedding Documents & Preparation." www. usna.edu/Chapel/_files/documents/information/Catholic%20Wedding%20 Info%20Aug%202013.pdf.

U.S. Navy records of Lisa Marie Nowak.

Yearbook pages on Lisa Caputo and Richard Nowak. 1985.

Secondary Sources

Aldrin, Buzz. *Magnificent Desolation: The Long Journey Home from the Moon.* New York: Harmony, 2009.

———. *No Dream Is Too High: Life Lessons from a Man Who Walked on the Moon.* Washington, D.C.: National Geographic, 2016.

Allen, Buckie. *Lawyers of Orange County.* Self-published, 2011.

Anderson, Clayton, "Does The International Space Station Have an On-Call Doctor?" gizmodo.com. October 21, 2015.

"Astronauts Enjoy Slightly More Relaxed Day." Space Place webpage. CBS News. July 9, 2006.

Austin Peterson, Liz. "Astronaut Mourns His Mother from Orbit." Associated Press, December 20, 2007.

Baker, Mike. "NASA Astronaut Anne McClain Accused by Spouse of Crime in Space." *New York Times,* August 23, 2019.

Blackwell Landon, Lauren, Kelly Slack, and Jamie Barrett. "Teamwork and Collaboration in Long-Duration Space Missions: Going to Extremes." *American Psychologist* 73, no. 4 (January 24, 2017).

Bracther, Drew. "Lisa Nowak: An Astronaut Reaching for the Stars." *Washingtonian,* January 1, 2007.

Bradley, Ed. "First Man: Neil Armstrong Interview." *60 Minutes.* CBS. August 25, 2012.

Collins, Michael. PBS *NewsHour* interview. July 17, 2019.

Como, Chris. "Astronaut Love Triangle Victim." *20/20.* February 18, 2011.

Epstein, Robert. "Buzz Aldrin: Down to Earth." *Psychology Today,* May 1, 2001.

Fleeman, Michael. "Nowak Apologizes." *People,* August 24, 2007.

Gifford, Gabrielle, and Mark Kelly. *Gabby: A Story of Courage and Hope.* New York. Scribner, 2011.

Gwynne, S. C. "Lust in Space." *Texas Monthly,* May 2007.

Harnden, Toby. "Searchers Stumble on Human Remains." *Telegraph* (London), February 3, 2003.

Hickam, Homer. "What Makes an Astronaut Crack?" *Los Angeles Times,* February 9, 2007.

Hidalgo Whitesides, Loretta. "Even Astronauts Commit Suicide: A Tribute to a Friend and a Plea." *Wired,* September 28, 2007.

Interview of STS-121 Shuttle Crew. *USA Today,* July 12, 2006.

Interview with Jonathan Clark, widower of astronaut Laurel Clark, space shuttle crew surgeon and chief of the Medical Operations Branch at Johnson Space Center. *Nancy Grace Show.* February 2007.

Kalb, Claudia. "The Lisa Nowak I Knew." *Newsweek,* November 12, 2009.

Lenhart, Jennifer, "Astronaut from Rockville Keeps Her Eyes on Space, Heart on Earth." *Washington Post,* July 4, 2006.

Letterman, David. "Top 10 Signs an Astronaut Is Trying to Kill You." *CBS Late Show.* February 7, 2007.

Levine, David. "Can Humans Live in Space without Going Crazy?" *Astronomy,* August 14, 2018.

"Nextfest: Tomorrow's Technology Today." *Wired,* September 2006.

"Other Man's Mother-in-Law: Nowak Broke up My Daughter's Marriage." FOX News. February 8, 2007.

McLaughlin, Moira. "Shuttle Astronaut Visits Stone Ridge." *Catholic Monthly,* October 26, 2006.

Menon, Anil, Shannon Moynihan, Kathleen Garcia, and Ahot Sargsyan. "How NASA Uses Telemedicine to Care for Astronauts in Space." *Harvard Business Review,* July 6, 2017.

Morris, Rodger. "Flight Training at Pensacola Naval Air Station." *BlackFive* (blog). 2008. www.blackfive.net.

Nelson, Melissa. "Florida Naval Base to Graduate Final Class." Associated Press, February 20, 2008.

Oefelein, William. Biography on Adventure Write website. 2008.

———. Launch blog. December 7–22, 2006. www.nasa.gov/mission_pages/shuttle/shuttlemissions/sts116/oefelein_blog.html.

———. "Left for Dead." Forget Me Not Foundation: Anchorage, AK, 2005. www.forgetmenotmission.com/soul_shaking_grief/soul_shaking_grief.pdf.

Reagan, President Ronald. Speech to Naval Academy graduates. May 22, 1985. www.youtube.com.

Ruff, Capt. (USAF) George, and Capt. (USAF) Edwin Levy. "Psychiatric Evaluation of Candidates for Space Flight." *American Journal of Psychiatry* 116, no. 5 (November 1959).

Ryan, Sarah. "Behind the Scenes: Canadian Military Trains for Wilderness Survival." January 21, 2017. globalnews.ca.

Santy, Patricia, M.D. *Choosing the Right Stuff: The Psychological Selection of Astronauts and Cosmonauts.* Westport, CT: Praeger, 1994.

Shipman, Collen. Biography on Adventure Write website. 2008.

"Space Shuttle Disaster." *Nova* (television show). June 22, 2011.

Tresniowski, Alex, Rose Ellen O'Conner, Jeff Truesdale, Steve Helling, Alicia Dennis, Wendy Grossman, and John Perra. "Out of This World." *People* magazine, February 19, 2007.

Tunseth, Matt, "Alaska Astronaut Inspires Kenai Peninsula Youngsters." *Peninsula Clarion,* July 10, 2003.

Turner, Allan, "Astronaut in Middle Looked, Acted Part of Space Pioneer." *Houston Chronicle,* February 6, 2007.

"U.S. Astronaut Lisa Nowak, a Short Biography." February 7, 2007. newscientist.com.

Watson, Traci. "Astronaut Posts Bail after Being Charged with Attempted First-Degree Murder." February 7, 2007.

———. "Astronauts Wield Giant Helping Hand." *USA Today,* July 10, 2006.

———. "Charges and Intrigue a New World for Shuttle Astronaut." With Andrea Stone. February 7, 2007.

FLORIDA TODAY STORIES, 2002–2010

"Emails Reveal Nowak Friendship." December 14, 2007.

"Ex-Astronaut Lisa Nowak's Navy Career Over." August 20, 2010.

Halvorson, Todd. "Star-Spangled Start." July 5, 2006.

———. "It's Moving Day in Space." July 7, 2006.

———. "Thumbs up for Boom." July 9, 2006.

———. "Repairs Allow Work on Station to Resume." July 11, 2006.

———. "Astronauts Enjoy Day off in Orbit." July 14, 2006.

———. "Shuttle to Leave Station Today, No Damage Found on Shuttle." July 15, 2006.

———. "Astronauts Must Undergo Rigorous Testing." February 7, 2007.

———. "Screenings under Scrutiny." February 8, 2007.

———. "Restaurant Plans Nowak Benefit." February 9, 2007.

———. "Fuhrman Weighs in on Astronaut's Case." February 10, 2007.

———. "Note Shows Relationship Heating Up." March 7, 2007.

———. "NASA Done with Oefelein. May 26, 2007.

Halvorson, Todd, and Chris Kridler. "The Crew." July 4, 2006.

———. "Astronauts Share Space Stories." July 12, 2006.

———. "Tidbits and Trivia from Space." July 19, 2006.

Kelly, John. "E-mails May Have Fueled Rage." March 7, 2007.

———. "NASA Dismisses Troubled Nowak." March 8, 2007.

———. "Colleagues Suspected Love Affair." May 2, 2007.

Kelly, John, and Chris Kridler. "Columbia Lost." February 1, 2003.

Kelly, John, and Kimberly C. Moore. "For NASA's Nowak, a Stunning Descent." February 7, 2007.

———. "NASA to Review Mental Health of Astronauts." February 9, 2007.

Kridler, Chris. Profile of Pilot William McCool. January 2003.

———. "It All Leads up to This." October 6, 2002.

———. "Nowak Won't Overlook Small Details of Mission." June 27, 2006.

———. "Cleared to Come Home." July 17, 2006.

———. "Nowak Gave No Hint of Turmoil." February 7, 2007.

Moore, Kimberly C. "Astronaut Brady Dies in Apparent Suicide." July 30, 2006.

———. "Navy Letting Nowak Case 'Run Its Course.'" February 17, 2007.

———. "Nowak Begins Trial July 20." March 23, 2007.

———. "More Evidence Released in Nowak Case." April 11, 2007.

———. "Nowak's Car Contained Bondage Photos." April 11, 2007.

———. "Items Show Intent, Police Say." April 12, 2007.

———. "Space Shuttle Launch Experience Opening." May 26, 2007.

———. "Shipman Gives Deposition in Nowak Case." June 23, 2007.

———. "Nowak's Attorney Denies Diapers." June 30, 2007.

———. "Video Shows Jailed Nowak." August 11, 2007.

———. "Media Hits Orlando for Nowak Hearing." August 24, 2007.

———. "Nowak Sorry She Frightened Shipman." August 25, 2007.

———. "Nowak Claims Insanity." August 29, 2007.

———. "Judge: Nowak Can Ditch Ankle Device." August 31, 2007.

———. "Navy Issues Nowak Order." September 13, 2007.

———. "Nowak: I Asked for an Attorney." September 20, 2007.

———. "Judge Tosses Evidence, Statements in Nowak Case." November 3, 2007.

———. "Nowak Evidence Disputed in Court." October 22, 2008.

———. "Evidence Found in Lisa Nowak's Car Will Be Allowed at Her Trial." December 6, 2008.

———. "No Trial, but Many Changes for Nowak." February 16, 2009.

———. "Nowak to Undergo Testing." April 2, 2009.

———. "Nowak Defense Blasts State." April 14, 2009.

———. "No Jail for Nowak." November 11, 2009.

———. "Career of Ex-Astronaut Nowak Now in Navy Hands." February 1, 2010.

———. "Military Hearing Held to Consider Punishment for Love-Triangle Astronaut." August 20, 2010.

Torres, John. "Nowak's Attorney Given Extension." May 5, 2007.

———. "Nowak OK with Sharing Oefelein." July 10, 2007.

———. "Ruling Delays Nowak Trial." January 17, 2008.

Index

O'Brien, Miles, 79–80
O'Conner, Bryan, 220
Oefelein, Billye, 50, 52, 174
Oefelein, Randall, 50, 52, 141
Oefelein, Randy, 50
Oefelein, Michaella Davis, 52, 57, 200
Oefelein, William (Billy): biography, 50–57; criticism of, 256; emails with Lisa Nowak, 62, 137–40, 145; Nowak love interest, 5–6, 58–62, 65, 87–89, 137–56, 200–202, 234–37; photographs, 129–31, 133; relationship with Shipman, 17–18, 30, 138–48, 200–202, 255, 261; STS-116 mission, 138–43
Onizuka, Ellison, 70
Operation Desert Storm, 67. *See also* Gulf War
Operation Southern Watch, 53–54
Oregon State University, 52
Orfinger, Judge Richard, 246
Orlando, 147, 158, 162, 173–74
Orlando International Airport, 162–66, 174
Orlando Police Department, 3, 175
Ostrom, Peter, 251

Paris, 134
Patrick Air Force Base, 16, 121
Patuxent River, Maryland, 4, 29
Payette, Julie, 58, 99
PBS *NewsHour,* 223
Peninsula Clarion, 51, 56
Pennsylvania State University, 120
Pensacola Naval Air Station, 25
Pentagon, 120
People, 262
Pesikoff, Richard, 204, 231–32, 234–37, 247
Picano, James, 209, 225
Pirate's Landing, 160
Pitcairn Island, 99
Pittsburgh, Pennsylvania, 120
Pittsfield, Massachusetts, 70
Pogue, William, 217
Polansky, Mark "Roman," 138, 140, 143

Portman, Natalie, 263
Price, Leontyne, 14
Pronk, John, 43
Psychology Today, 214
Pyrenees Mountains, 98

Quebec, Canada, 58
Quest Airlock, 102–3, 107

Ramon, Ilan, 37, 39
Reagan, President Ronald, 15, 26
Reeve, Wendell, 166
Reilly, Jim, 80
Reiter, Thomas, 65, 72–74, 77, 93, 97, 99, 101, 110, 116, 125, 128, 262; wife Consuela and sons Daniel and Sebastian, 97
REO Speedwagon, 12
Resnick, Judith, 70
Ride, Sally, 3, 12
Roan, Bernard, 240
Robber's Roost Canyon, Utah, 57
Rockville, Maryland, 4–5, 10, 13
Rominger, Kent, 75, 77
Royal Air Force Cadet Program, 71
Ruff, George, 210, 217
Ryan, Timothy, 165

Saint Arnold's Brewery, 147
Salzburg, Austria, 121
Santy, Dr. Patricia, 206, 211–13, 215–17, 220–22, 229, 256–57
Schneider, Mike, 191, 197
Sellers, Piers, 65, 71–72, 77, 97, 101–12, 115–16, 119, 125, 128–29, 158, 249, 262; Nowak's friend, 158, 249; STS-121 mission, 93–119; wife Mandy, 108
Senate Select Committee on Presidential Campaign Activities, 14
Severn River, 21
Seward, Alaska, 142
Shannon, John, 79, 81, 215–17
Shatner, William, 134
Shaw, Brewster, 219–20
Shepherd, Alan, 19

KIMBERLY C. MOORE is an investigative reporter in central Florida. She spent time as an anchor and reporter in Israel during the first Gulf War and has covered the U.S. Congress and the White House in Washington, D.C., along with the contested 2000 presidential election in Tallahassee, and multiple space shuttle launches. She has earned numerous journalism awards.